Process Development Symposium 2015

Emerging Trends in Process Research

Houston, Texas, USA
2 – 4 June 2015

ISBN: 978-1-5108-1575-9

Printed from e-media with permission by:

Curran Associates, Inc.
57 Morehouse Lane
Red Hook, NY 12571

proceedings.com

Some format issues inherent in the e-media version may also appear in this print version.

Printed by Curran Associates, Inc. (2015)

For permission requests, please contact AIChE
at the address below.

AIChE
120 Wall Street, FL 23
New York, NY 10005-4020

Phone: (800) 242-4363
Fax: (203) 775-5177

www.aiche.org

Additional copies of this publication are available from:

Curran Associates, Inc.
57 Morehouse Lane
Red Hook, NY 12571 USA
Phone: 845-758-0400
Fax: 845-758-2634
Email: curran@proceedings.com
Web: www.proceedings.com

TABLE OF CONTENTS

Some Challenges and Opportunities for Process Development in the 21st Century

West, D. H., Dow Chemical Company

The basic chemicals and plastics industry is built on a simple idea: convert a cheap feedstock into something more valuable. Because of recent trends in oil production, we have seen some interesting changes in feedstock availability and price. These have caused a major shift in the economy of producing basic chemicals and polymers; restoring competitiveness in the US, while reducing it in Europe. This situation will likely persist for the near future. Nothing else effects the profitability of the chemical industry more than the feedstock. So what can be done in the way of process development? Many of the transformative process innovations of the 20th century were driven by innovations in catalysis: the Haber-Bosch process for ammonia; zeolite-Y and fluidized catalytic cracking for gasoline; methane reforming; Fischer-Tropsch chemistry; Ziegler-Natta, and more recently, metallocene catalysts for polyolefins. What is left to do? I will discuss some process development opportunities that remain and some approaches to those challenges that we are exploring at SABIC.

PDS Process Development Symposium

AIChE The Global Home of Chemical Engineers

Modeling and Optimization of Multi-layer Structured Catalysts for Selective Elimination of NO_x and NH_3

Mike Harold *

University of Houston,

Dept. of Chemical & Biomolecular Engineering,

Houston, TX (USA)

** mharold@uh.edu*

1

PROCESS DEVELOPMENT SYMPOSIUM, JUNE 2-4, 2015, HOUSTON, TX

PDS Process Development Symposium

AIChE The Global Home of Chemical Engineers

Acknowledgements

- **Collaborators**
  - Prof. Vemuri Balakotaiah, Prof. Dan Luss
  - Current Students: Sachi Shrestha
  - Former Students: Pranit Metkar (DuPont), Bijesh Shakya (Johnson Matthey)
  - Dr. Krishna Kamasamadram, Dr. Alexsey Yezerets (Cummins)
- **Funding**
  - DOE Vehicle Technologies Program
  - National Science Foundation
  - Cummins

2

PROCESS DEVELOPMENT SYMPOSIUM, JUNE 2-4, 2015, HOUSTON, TX

Motivation

AIChE
The Global Home of Chemical Engineers

http://solar.calfinder.com/blog/wp-content/uploads/2009/12/toxic-city-houston.jpg
http://images.google.com/imgres?imgurl=http://www.utexas.edu/research/ceer/texaqs/images/downtown_view2

3

PROCESS DEVELOPMENT SYMPOSIUM, JUNE 2-4, 2015, HOUSTON, TX

Three-Way Catalytic Converter

AIChE
The Global Home of Chemical Engineers

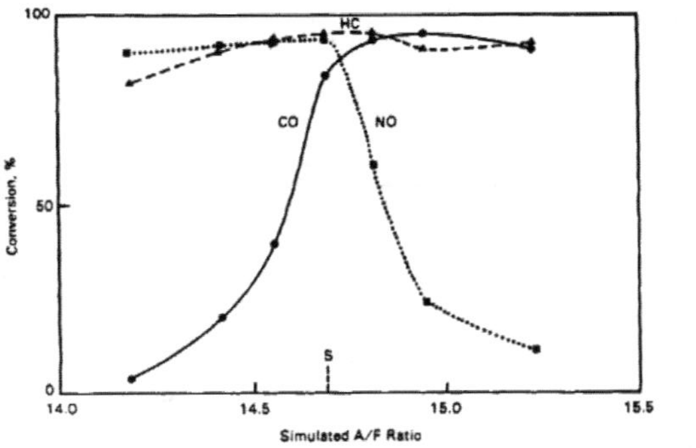

FIG. 1. Principle of 3-way exhaust emission control. Catalyst efficiencies measured in the laboratory with a steady feed stream composition at various simulated air/fuel ratios. Catalyst: 0.042 wt% Pt/0.018 wt% Rh/alumina [11].

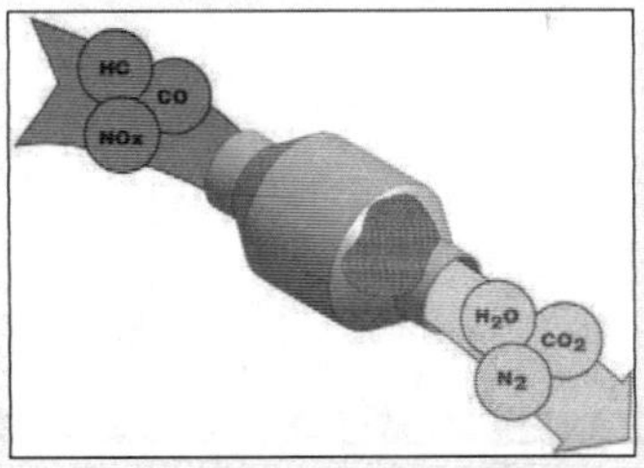

$Pd/CeO_2/Al_2O_3$

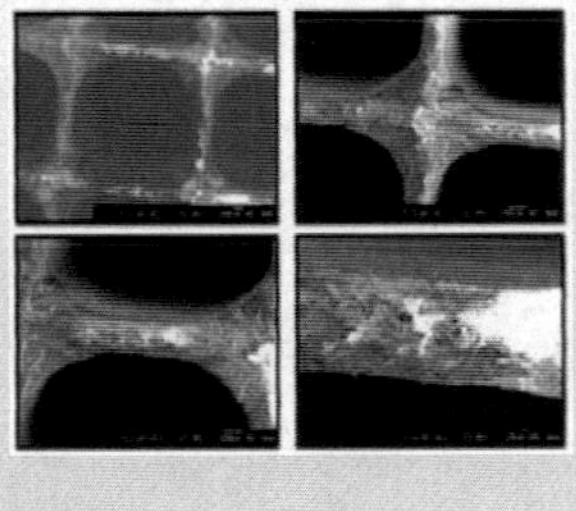

Ref.: K. C. Taylor (1993): Nitric Oxide Catalysis in Automotive Exhaust Systems, Catalysis Reviews: Science and Engineering, 35:4, 457-481

4

PROCESS DEVELOPMENT SYMPOSIUM, JUNE 2-4, 2015, HOUSTON, TX

PDS Process Development Symposium

AIChE The Global Home of Chemical Engineers

Requirements of Catalytic Converter

"The vehicle converter is a compact catalytic reactor that is started up and shut down several times a day, has a feed flow rate and a composition that vary by up to an order of magnitude in a period of seconds, has typical contact times on the order of 10s of milliseconds, must operate at temperatures from ambient up to 1000 °C, has product yields measured from start-up with no allowance for start-up losses, and yet must achieve over 99% conversion of the primary pollutants. To top it off, the reactor must maintain its performance for 1000s of hours while being operated by unskilled personnel (adapted from McCabe [1])."

Ref: Harold, M.P., "NOx Storage and Reduction in Lean Burn Vehicle Emission Control: A Catalytic Engineer's Playground," Current Opinion in Chemical Engineering, **1**, 1-9 (2012).

PROCESS DEVELOPMENT SYMPOSIUM, JUNE 2-4, 2015, HOUSTON, TX

5

PDS Process Development Symposium

AIChE The Global Home of Chemical Engineers

Challenges

- Conversion: Increasingly stringent emissions standards
- Control: Close coupling of emissions system and engine
- Cost: Precious metals common but are expensive ~$750 - $1,500/ounce

PROCESS DEVELOPMENT SYMPOSIUM, JUNE 2-4, 2015, HOUSTON, TX

6

Process Development Symposium

AIChE The Global Home of Chemical Engineers

US Heavy Duty Diesel Emissions Standards

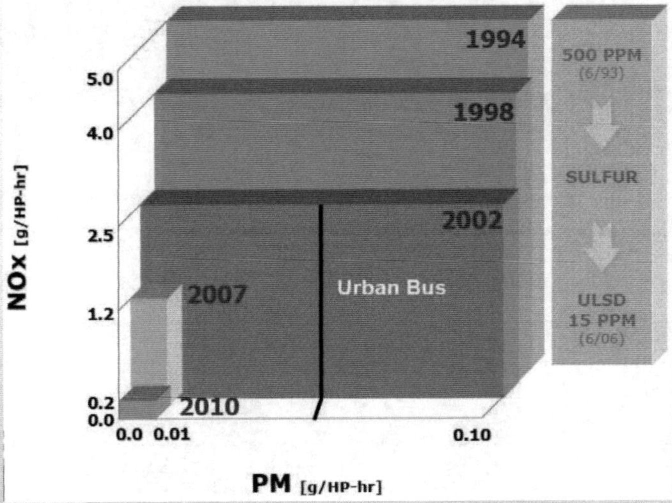

PROCESS DEVELOPMENT SYMPOSIUM, JUNE 2-4, 2015, HOUSTON, TX

7

Clean Diesel Emission Technology

Evolving **New Fuels** →

→ **Diesel Oxidation Catalyst** → **Diesel Particulate Filter** → **NOx Reduction Reactor** → **NH_3 Slip Catalyst** → *Clean Exhaust*

Abatement: **VOCs & CO** | **Soot & Solids** | **NO & NO_2**

Status: *Implemented* | *Maturing* | *Early commercialization*

A chemical plant on wheels ....

AIChE
The Global Home of Chemical Engineers

Storage & Reaction on Multi-Functional Catalysts in Exhaust Aftertreatment

Method	Application	Reaction	Catalyst	Stored Species
TWC	Spark-ignited gasoline	H_2/CO/HC + O_2	Pt/Pd/Rh/CeO_2/Al_2O_3	O_2
DOC	Diesel	CO/HC + O_2	Pt/Pd/zeolite-β/Al_2O_3	High MW HC
DPF	Diesel	C + O_2/NO_2	Pt/cordierite	Soot
NSR	Lean burn, Diesel	H_2/CO/HC + NOx	Pt/Rh/BaO/CeO_2/Al_2O_3	NOx
SCR	Diesel	NH_3 + NO +NO_2	Cu or Fe/zeolite	NH_3
ASC	Diesel	NH_3 + O_2	Cu/zeolite + Pt/Al_2O_3	NH_3

PROCESS DEVELOPMENT SYMPOSIUM, JUNE 2-4, 2015, HOUSTON, TX

9

AIChE
The Global Home of Chemical Engineers

Storage & Reaction on Multi-Functional Catalysts in Exhaust Aftertreatment

Method	Application	Reaction	Catalyst	Stored Species
TWC	Spark-ignited gasoline	H_2/CO/HC + O_2	Pt/Pd/Rh/CeO_2/Al_2O_3	O_2
DOC	Diesel	CO/HC + O_2	Pt/Pd/zeolite-β/Al_2O_3	High MW HC
DPF	Diesel	C + O_2/NO_2	Pt/cordierite	Soot
NSR	Lean burn, Diesel	H_2/CO/HC + NOx	Pt/Rh/BaO/CeO_2/Al_2O_3	NOx
SCR	Diesel	NH_3 + NO +NO_2	Cu or Fe/zeolite	NH_3
ASC	Diesel	NH_3 + O_2	Cu/zeolite + Pt/Al_2O_3	NH_3

PROCESS DEVELOPMENT SYMPOSIUM, JUNE 2-4, 2015, HOUSTON, TX

10

PDS Process Development Symposium

AIChE The Global Home of Chemical Engineers

Architectures of Emission Catalysts

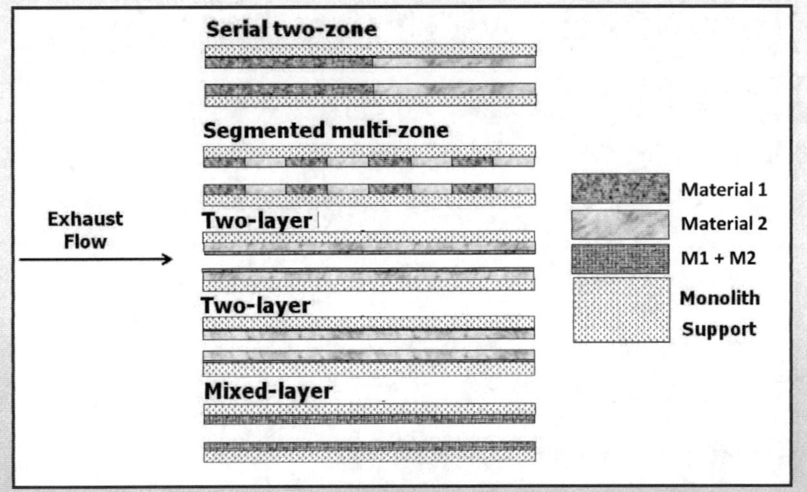

(adapted from Gandhi et al., US Patent, 2007)

PROCESS DEVELOPMENT SYMPOSIUM, JUNE 2-4, 2015, HOUSTON, TX

11

PDS Process Development Symposium

AIChE The Global Home of Chemical Engineers

Architectures of Emission Catalysts

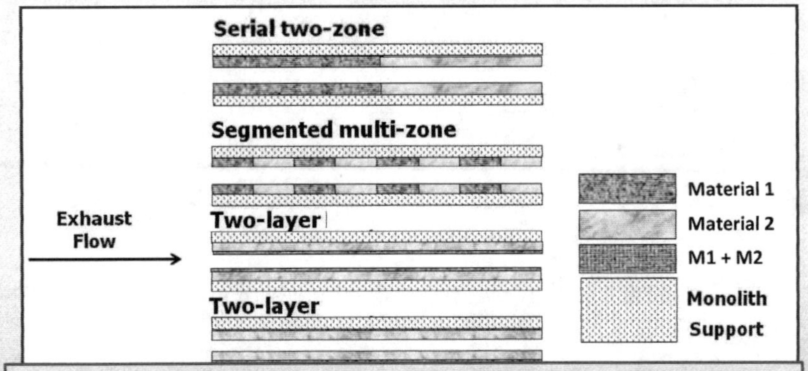

Conduct experiments & modeling to advance understanding & application of reaction & transport interactions in dual-layer emission catalysts

PROCESS DEVELOPMENT SYMPOSIUM, JUNE 2-4, 2015, HOUSTON, TX

12

I Hope to Show You That:

- Lean NOx catalysts inherently multi-functional
- Layering of two catalyst functions can enhance conversion and/or selectivity
- Focused experiments combined with modeling provides proof of concept, understanding, optimized catalysts, & ideas for improvement

13

PROCESS DEVELOPMENT SYMPOSIUM, JUNE 2-4, 2015, HOUSTON, TX

Monolith Reactor Model

Multiple Length Scales

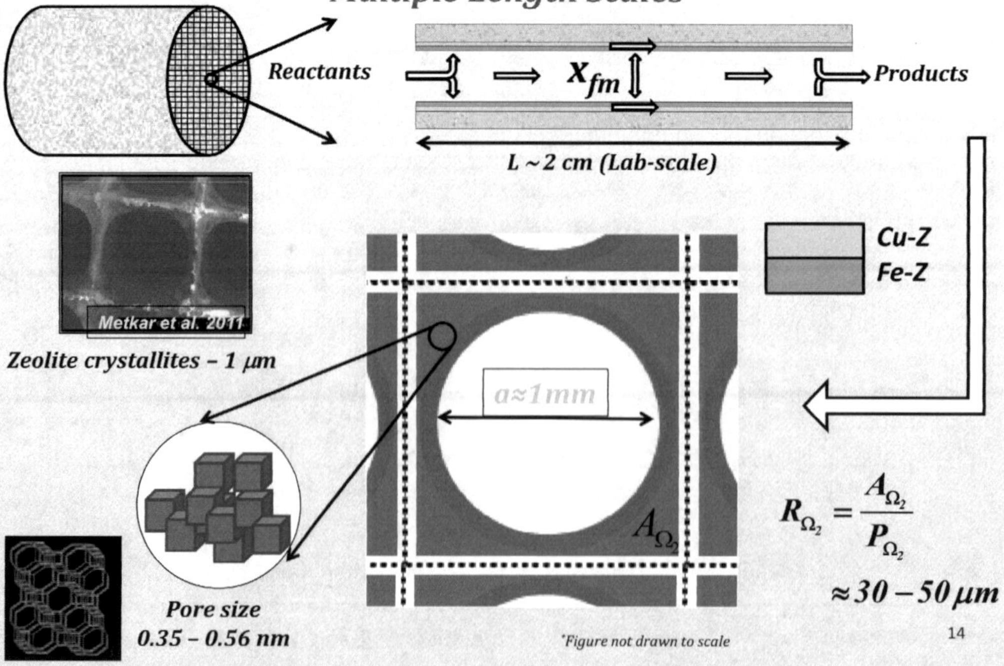

14

AIChE
The Global Home of Chemical Engineers

Two Examples

- Fe + Cu SCR Catalyst
- Ammonia Slip Catalyst

complexity

15

AIChE
The Global Home of Chemical Engineers

Two Examples

Fe + Cu SCR Catalyst

Exploitation of diffusion limitations

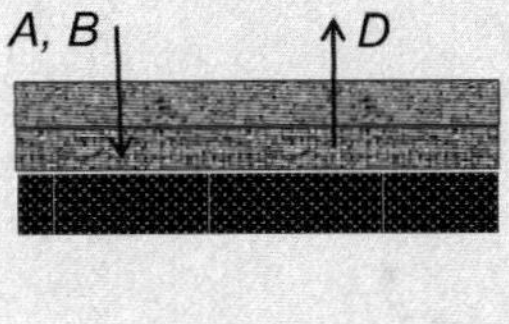

(Low T)
A + B → D

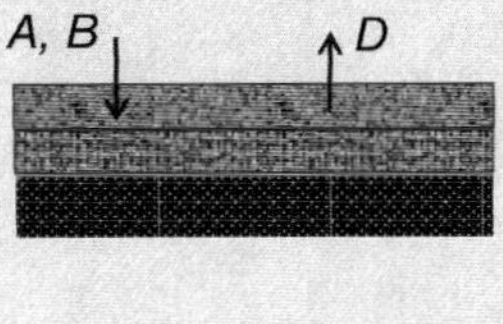

(High T)
A + B → D
A + B → E

16

Selective Catalytic Reduction (SCR)

Standard SCR (Desired):

$4\ NH_3 + 4\ NO + O_2 \rightarrow 4\ N_2 + 6\ H_2O$

Fast SCR (Desired):

$2\ NH_3 + NO + NO_2 \rightarrow 2\ N_2 + 3H_2O$

Ammonia Oxidation (Undesired):

$4\ NH_3 + 3\ O_2 \rightarrow 2\ N_2 + 6H_2O$

Catalysts:

Cu- & Fe-exchanged zeolites
(CHA, ZSM-5, BEA)

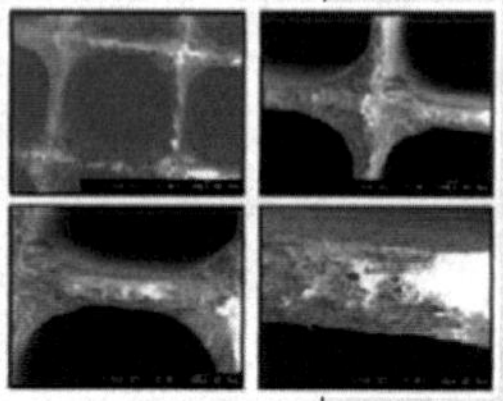

SCR Catalyst Comparison

- Cu-based:
  - More active
  - High temp. NH_3 oxidation
- Fe-based:
  - High temp. activity
  - NH_3 inhibition

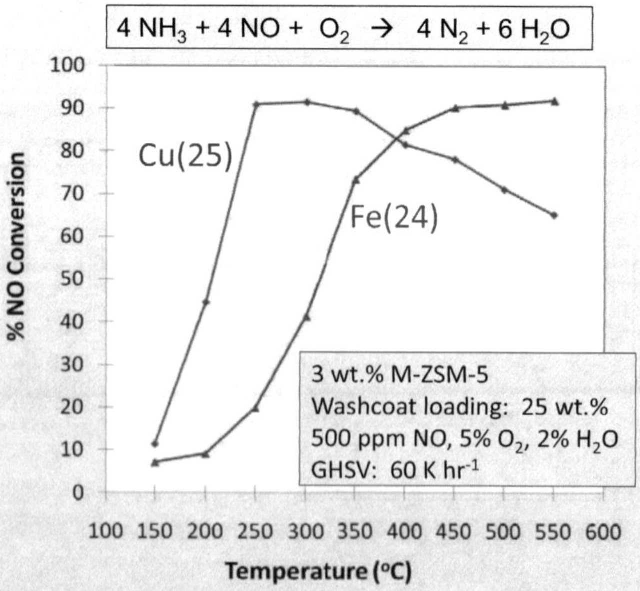

Metkar P.S., Harold M.P, Balakotaiah, V., Appl. Catal. B 111-112, 67-80 (2012)
Girard, J., C. Cavataio, R. Snow, C. Lambert, SAE Paper 2008-01-1185 (2008)

Dual Layer SCR Catalyst

$4\ NH_3 + 4\ NO + O_2 \rightarrow 4\ N_2 + 6\ H_2O$

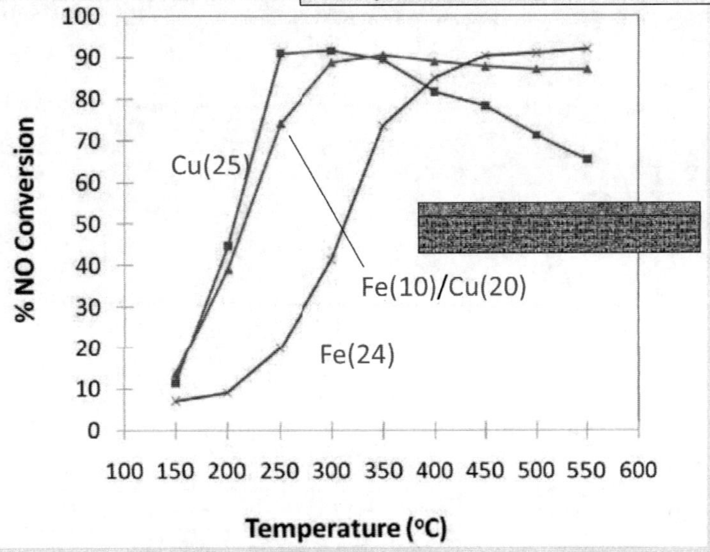

Metkar P.S., Harold M.P, Balakotaiah, V., Appl. Catal. B 111-112, 67-80 (2012)
Metkar P.S., Harold M.P, Balakotaiah, V., Patent Pending (2011)

Dual Layer SCR Catalyst: Concept

$4\ NH_3 + 4\ NO + O_2 \rightarrow 4\ N_2 + 6\ H_2O$

@ low temperature

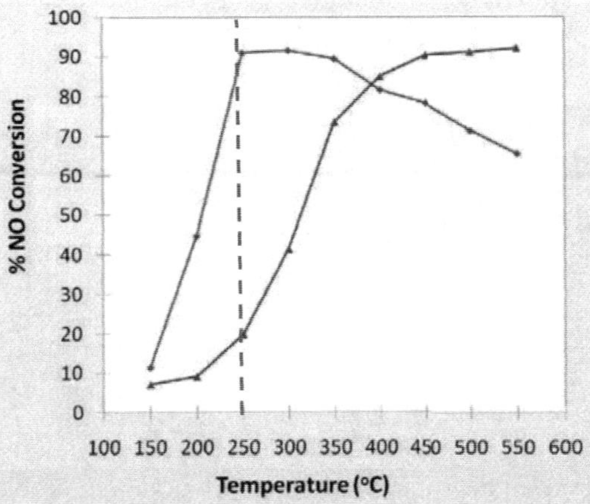

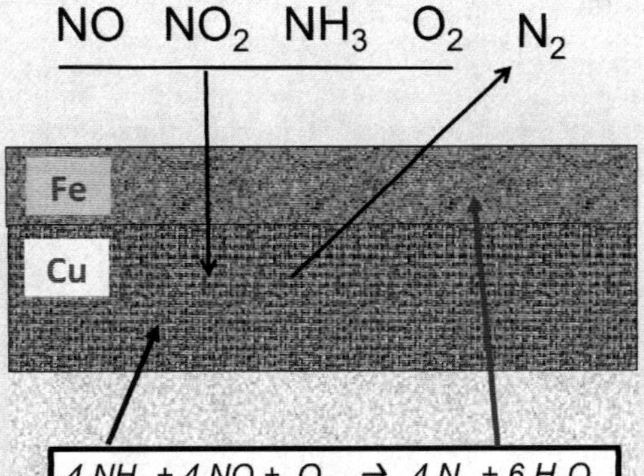

occurs in Cu layer; minimized in Fe layer

Dual Layer SCR Catalyst: Concept

$4\,NH_3 + 4\,NO + O_2 \rightarrow 4\,N_2 + 6\,H_2O$

@ high temperature

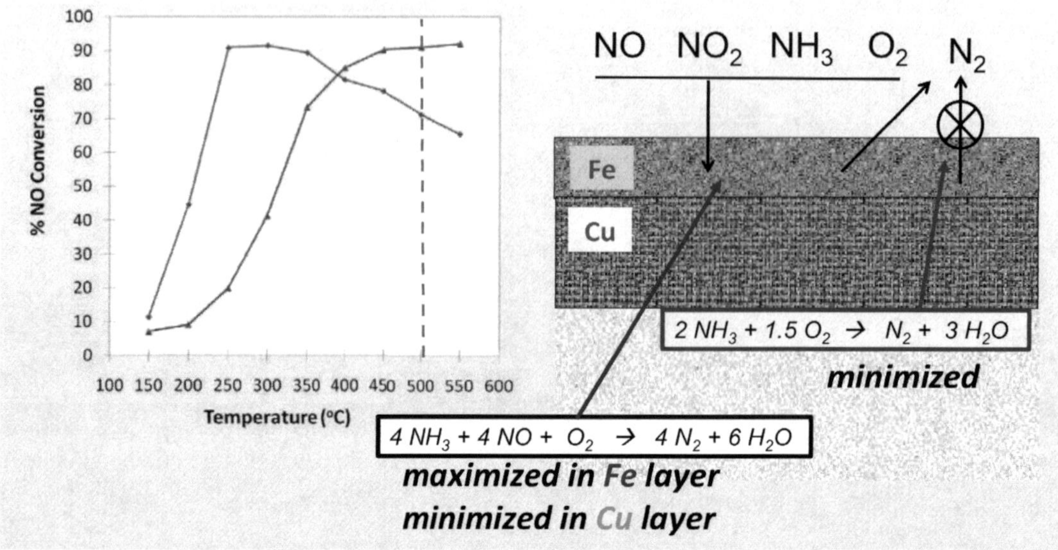

SCR Reactor Model

Multiple Length Scales

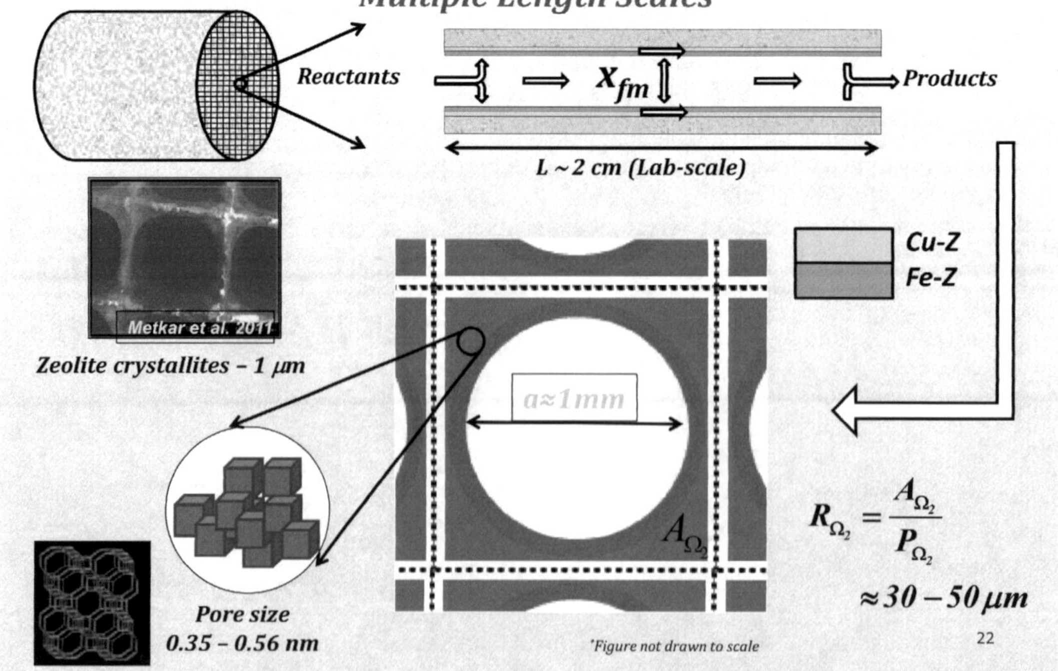

$$R_{\Omega_2} = \frac{A_{\Omega_2}}{P_{\Omega_2}}$$

$$\approx 30 - 50\,\mu m$$

22

Model Development

- 1 + 1 D Model
  - Shakya, B. M. et al. *Chem. Eng. J.*, 237, 109-122 (2014).

Fluid Phase:

$$\frac{\partial X_{jm}}{\partial t} = -\langle u \rangle \frac{\partial X_{jm}}{\partial z} - \frac{k_{c(j,z)}}{R_{\Omega 1}}\left(X_{jm} - X_{jwc}\big|_{y=0}\right).$$

Solid Phase:

$$\varepsilon_{wc}\frac{\partial X_{jwc}}{\partial t} = \frac{1}{C_{Tm}}\left(\sum \upsilon_j R_v(T,\theta,X_{jwc})\right) + De_j \frac{\partial^2 X_{jwc}}{\partial y^2}.$$

Adsorbed Species:

$$C_{st}\frac{\partial \theta_{js}}{\partial t} = \sum \upsilon_j R_v(T,\theta,X_{jwc}).$$

Initial & Boundary Conditions:

$X_{jm}(z) = X_{jm}^{o}(z)$ at $t = 0$

$X_{jm}(t) = X_{jm}^{in}(z)$ at $z = 0$

$\frac{\partial X_{jwc}}{\partial y} = 0$ at $y = R_{\Omega 2}$

$k_{cj}\left(X_{jm} - X_{jwc}\big|_{y=0}\right) = -D_{ej}\frac{\partial X_{jwc}}{\partial y}$ at $y = 0$

$k_{cj} = \frac{4R_{\Omega 1}}{Sh * D_{fj}}, \quad D_{e,j} = \frac{\varepsilon_{wc}}{\tau_{wc}} D_{f,j}$

- SCR Reaction kinetics
  - Metkar, P. S. et al. *Chem. Eng. Sci.*, 87, 51-66 (2013)
  - Adsorption of NH_3 in single site 'S' represents Bronsted acid site
  - Accounts for NH_3 inhibition of Std. SCR through NH_3 sorption on SCR sites, Lewis site

SCR‡ – Reactions & Kinetics

	Reaction	Rate Expression
R1	NH_3 + S ↔ NH_3-S	$k_{1f}X_{NH_3}\theta_v - k_{1b}\theta_{NH_3}$
R2	2NH_3-S + 1.5O_2 → N_2 + 3H_2O + 2S	$k_{2f}X_{O_2}\theta_{NH_3}$
R3	NO + ½ O_2 ↔ NO_2	$k_{3f}X^{1/2}_{O_2}X_{NO} - k_{3b}X_{NO_2}$
R4	4NH_3-S + 4NO + O_2 → 4N_2 + 6H_2O + 4S	$\frac{k_{4f}X_{NO}\theta_{NH_3}}{1+K^{*}X_{NH_3}}$
R5	2NH_3-S + NO + NO_2 → 2N_2 + 3H_2O + 2S	$k_{5f}X_{NO}X_{NO_2}\theta_{NH_3}$
R6	4NH_3-S +3NO_2 → 3.5N_2 + 6H_2O + 4S	$k_{6f}X_{NO_2}\theta_{NH_3}$
R7	2NH_3-S + 2NO_2 → N_2 + NH_4NO_3 + H_2O + 2S	$k_{7f}X_{NO_2}\theta_{NH_3}$
R8	NH_4NO_3 → N_2O + 2H_2O	$k_{8f}X_{NH_4NO_3}$
R9	2N_2O → 2N_2 + O_2	$k_{9f}X_{N_2O}$
R10	2NH_3-S + 3N_2O → 4N_2 + 3H_2O + 2S	$k_{10f}X_{N_2O}\theta_{NH_3}$

Catalysts : Commercial Cu-CHA (ca. 3%) and Fe-ZSM5 (ca. 3%)

‡ P.S. Metkar et al., Chemical Engineering Science 87 (2013) 51–66

24

Dual Layer SCR Catalyst: Optimization

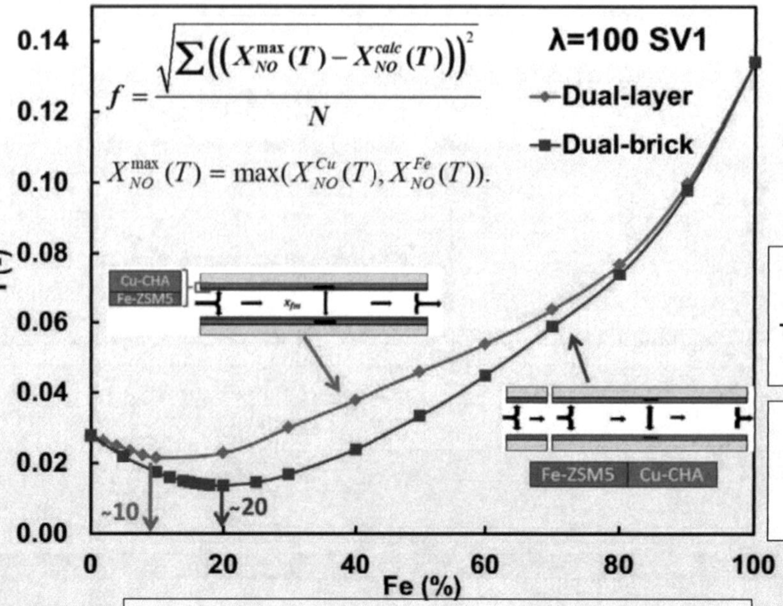

GHSV: 57,000 hr^{-1} (SV1); **Diffusivity Ratio:** $\lambda = D_m/D_e = 100$

Length = 2 cm, Washcoat Thickness = 50 μm, 34 channels

Feed: 500 ppm NO, 500 ppm NH_3, 5% O_2, 2% H_2O

Dual-brick catalysts give higher overall NO conversion than dual-layer

25

PDS Process Development Symposium

AIChE The Global Home of Chemical Engineers

Two Examples

Ammonia Slip Catalyst

Coupled reactions in adjacent layers

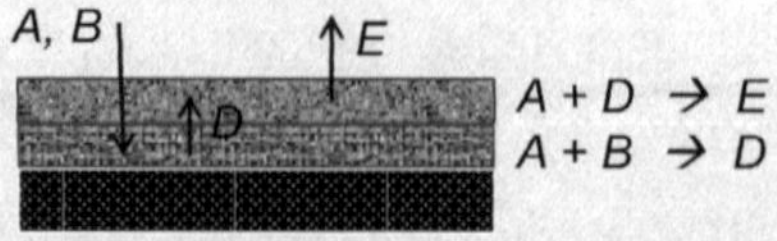

26

PROCESS DEVELOPMENT SYMPOSIUM, JUNE 2-4, 2015, HOUSTON, TX

ASC Concept

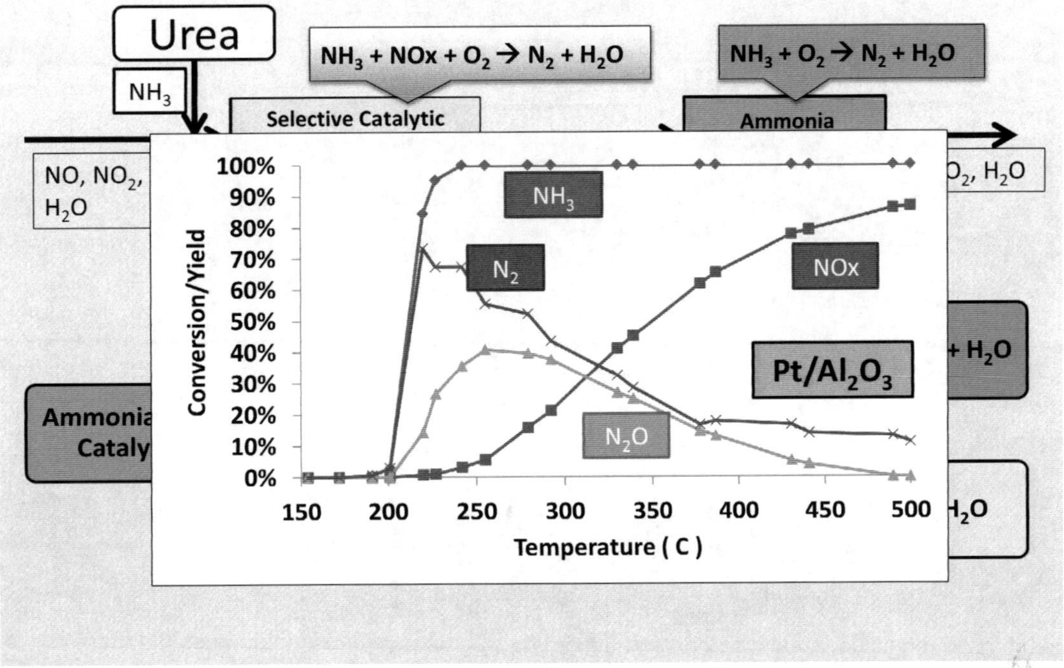

Dual Layer Ammonia Slip Catalyst

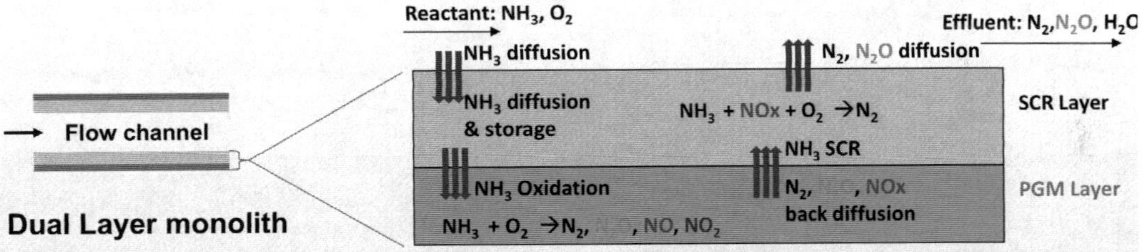

Reactions in SCR Layer

Standard SCR Reaction:

$4\ NH_3 + 4\ NO + O_2 \rightarrow 4\ N_2 + 6\ H_2O$

Fast SCR Reaction:

$2\ NH_3 + NO + NO_2 \rightarrow 2\ N_2 + 3\ H_2O$

NO_2 SCR Reaction:

$4\ NH_3 + 3\ NO_2 \rightarrow 3.5\ N_2 + 6\ H_2O$

Ammonia Oxidation

$4\ NH_3 + 3\ O_2 \rightarrow 2\ N_2 + 6\ H_2O$

Reactions in PGM Layer

$4\ NH_3 + 3\ O_2 \rightarrow 2\ N_2 + 6\ H_2O$

$2\ NH_3 + 2\ O_2 \rightarrow N_2O + 3\ H_2O$

$4\ NH_3 + 5\ O_2 \rightarrow 4\ NO + 6\ H_2O$

$4\ NH_3 + 7\ O_2 \rightarrow 4\ NO_2 + 6\ H_2O$

Adapted from Scheuer et al. *Applied Catalysis B: Environmental (2011)*

28

Demonstration of ASC Concept

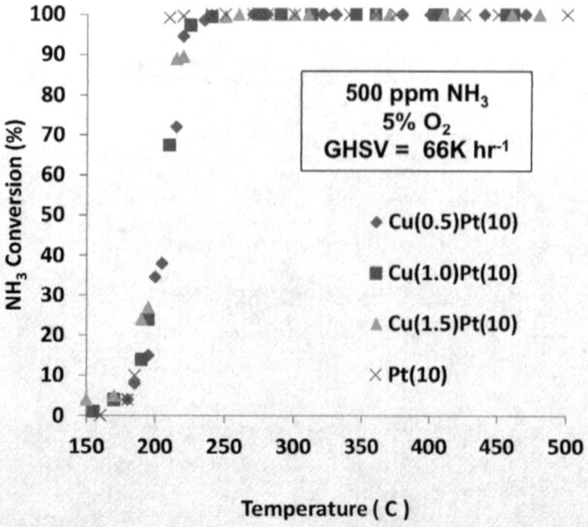

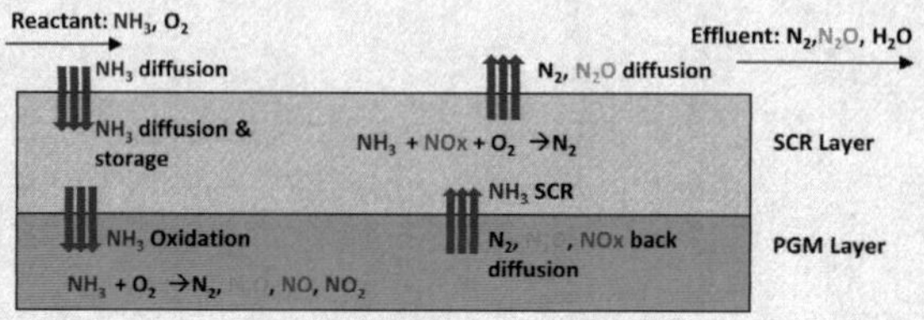

29

Demonstration of ASC Concept

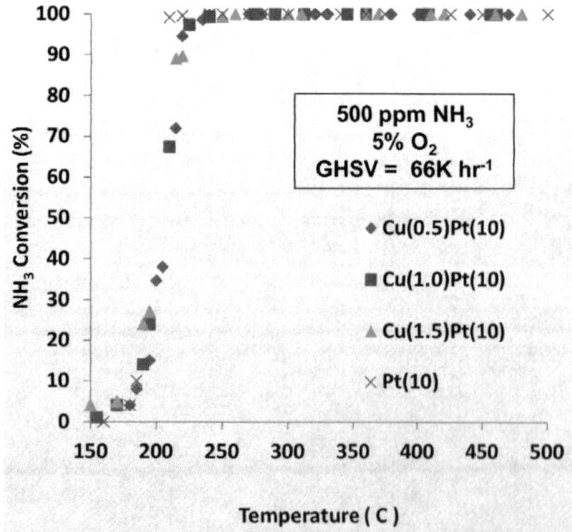

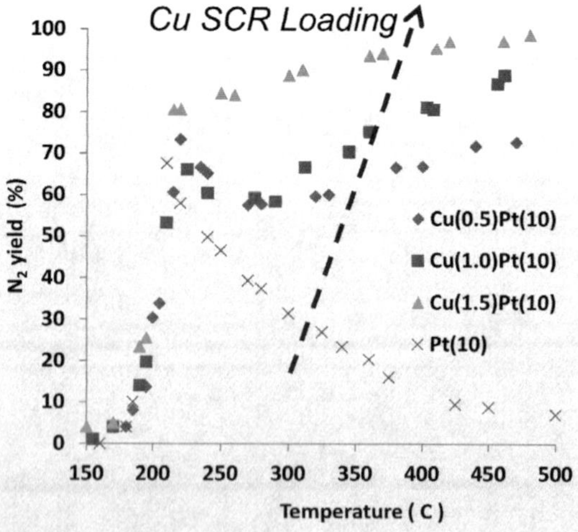

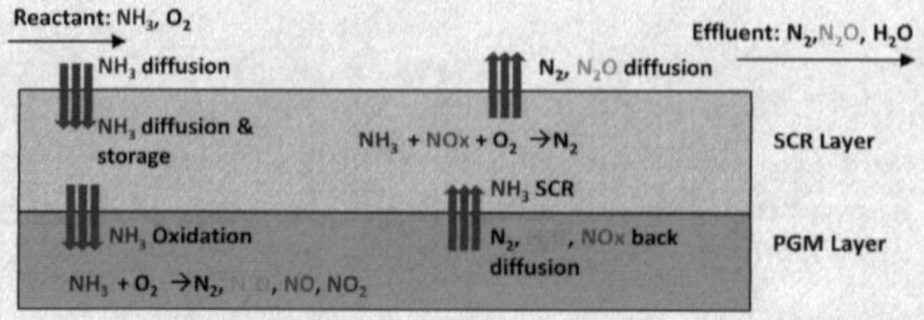

30

Model Development

- 1 + 1 D Model
  - Shakya, B. M. et al. *Chem. Eng. J.*, 237, 109-122 (2014).

Fluid Phase:

$$\frac{\partial X_{jm}}{\partial t} = -\langle u \rangle \frac{\partial X_{jm}}{\partial z} - \frac{k_{c(j,z)}}{R_{\Omega 1}}\left(X_{jm} - X_{jwc}\big|_{y=0}\right).$$

Solid Phase:

$$\varepsilon_{wc}\frac{\partial X_{jwc}}{\partial t} = \frac{1}{C_{Tm}}\left(\sum \upsilon_j R_v(T,\theta,X_{jwc})\right) + De\frac{\partial^2 X_{jwc}}{\partial y^2}$$

Adsorbed Species:

$$C_{st}\frac{\partial \theta_{js}}{\partial t} = \sum \upsilon_j R_v(T,\theta,X_{jwc}).$$

Initial & Boundary Conditions:

$X_{jm}(z) = X_{jm}^{o}(z)$	at $t = 0$
$X_{jm}(t) = X_{jm}^{in}(z)$	at $z = 0$
$\frac{\partial X_{jwc}}{\partial y} = 0$	at $y = R_{\Omega 2}$
$k_{cj}\left(X_{jm} - X_{jwc}\big|_{y=0}\right) = -D_{ej}\frac{\partial X_{jwc}}{\partial y}$	at $y = 0$

$$k_{cj} = \frac{4R_{\Omega 1}}{Sh * D_{fj}}, \qquad D_{e,j} = \frac{\varepsilon_{wc}}{\tau_{wc}} D_{f,j}$$

- SCR Reaction kinetics
  - Metkar, P. S. et al. *Chem. Eng. Sci.*, 87, 51-66 (2013)
  - Adsorption of NH_3 in single site 'S' represents Bronsted acid site
  - Accounts for NH_3 inhibition of Std. SCR through NH_3 sorption on SCR sites, Lewis site

- Pt Reaction kinetics
  - Kraenhnert, R. et al. *Chem. Eng. J., 137, 361-375* (2008)
  - Scheuer, A. et al. *Appl. Catal. B: Env.*, 111-112, 445-455 (2012)
  - 14 step microkinetic model
  - Top site 'b': NH_3 adsorption/desorption
  - Hollow site 'a': All other species
  - Entropic & enthalpic thermodynamic consistency for NO Oxidation
  - Optimized kinetic parameter

31

Ammonia Oxidation Kinetics

Reaction Number	Reaction	Rate Expression
Z1	$NH_3 + S \leftrightarrow NH_3\text{-}S$	$k_{1f}X_{NH_3}\theta_v - k_{1b}\theta_{NH_3}$
Z2	$2NH_3\text{-}S + 1.5O_2 \rightarrow N_2 + 3H_2O + 2S$	$k_{2f}X_{O_2}\theta_{NH_3}$
Z3	$NO + ½ O_2 \leftrightarrow NO_2$	$k_{3f}X_{O_2}^{\frac{1}{2}}X_{NO} - k_{3b}X_{NO_2}$
Z4	$4NH_3\text{-}S + 4NO + O_2 \rightarrow 4N_2 + 6H_2O + 4S$	$\frac{k_{4f}X_{NO}\theta_{NH_3}}{1 + K^{*}X_{NH_3}}$
Z5	$2NH_3\text{-}S + NO + NO_2 \rightarrow 2N_2 + 3H_2O + 2S$	$k_{5f}X_{NO}X_{NO_2}\theta_{NH_3}$
Z6	$4NH_3\text{-}S + 3NO_2 \rightarrow 3.5N_2 + 6H_2O + 4S$	$k_{6f}X_{NO_2}\theta_{NH_3}$
Z7	$2N_2O \rightarrow 2N_2 + O_2$	$k_{7f}X_{N_2O}$
Z8	$2NH_3\text{-}S + 3N_2O \rightarrow 4N_2 + 3H_2O + 2S$	$k_{8f}X_{N_2O}\theta_{NH_3}$

- Developed by Metkar et al. *Chemical Engineering Science* (2013)
- Adsorption of NH_3 in single site 'S' represents Bronsted acid site
- Accounts for NH_3 inhibition of Std. SCR through NH_3 sorption on SCR sites, Lewis site
- Ammonium Nitrate formation/decomposition neglected

32

Approach

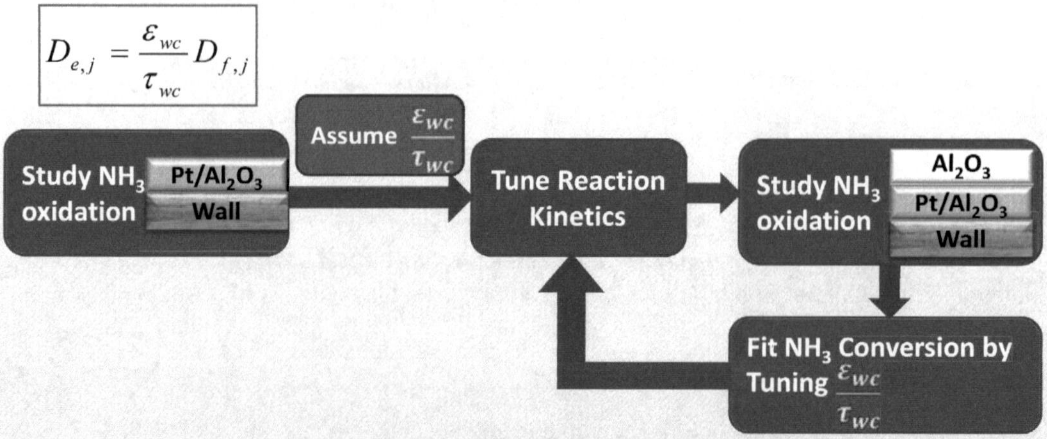

For SCR catalyst, similar technique to calculate $\frac{\varepsilon_{wc}}{\tau_{wc}}$

33

Diffusivity in Al_2O_3

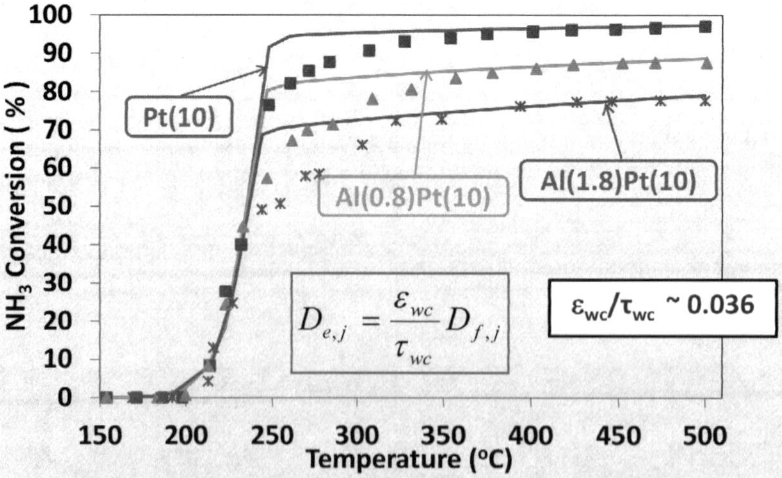

Feed: 500 ppm NH_3, 5% O_2, 2.5% H_2O, 2% CO_2, bal. Ar

Experimental: Symbols

Model: Lines

GHSV = 285k h^{-1}

34

Diffusivity in Na-ZSM-5

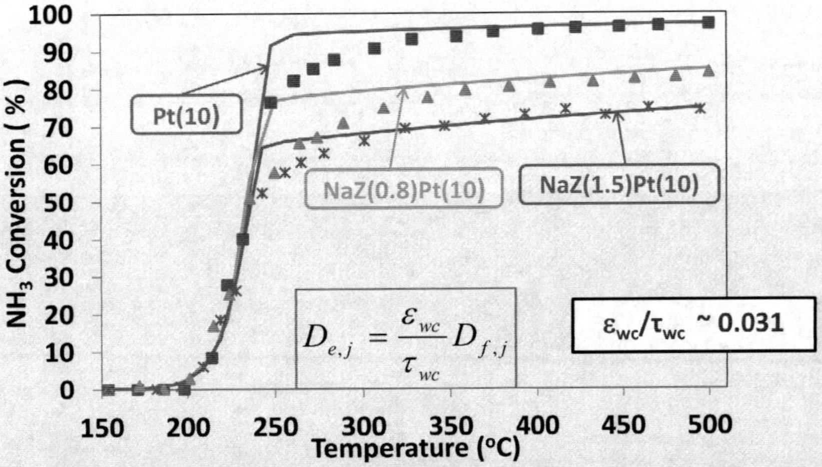

500 ppm NH_3, 5% O_2, 2.5% H_2O, 2% CO_2 , bal. Ar
Experimental: Symbols
Model: Lines

GHSV = 285k h^{-1}

35

Pt/Al_2O_3 Results

Catalyst	Pt Loading (g/ft^3 monolith)	Pt/Al_2O_3 Loading (g/in^3 monolith)	Pt Loading (g Pt per 100 g washcoat)
Pt(10)	10.5	1.4	0.46

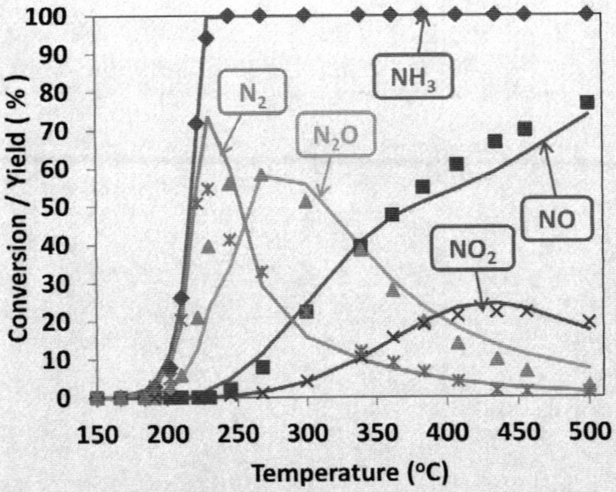

500 ppm NH_3, 5% O_2, 2.5% H_2O, 2% CO_2, bal. Ar
Experimental: Symbols
Model: Lines

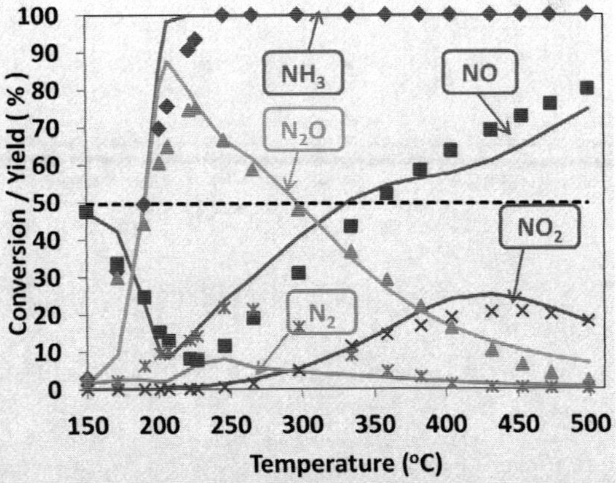

500 ppm NH_3, 500 ppm NO, 5% O_2, 2.5% H_2O, 2% CO_2, bal. Ar
Experimental: Symbols
Model: Lines

GHSV = 66k h^{-1}

36

Dual-layer vs. Mixed Catalyst

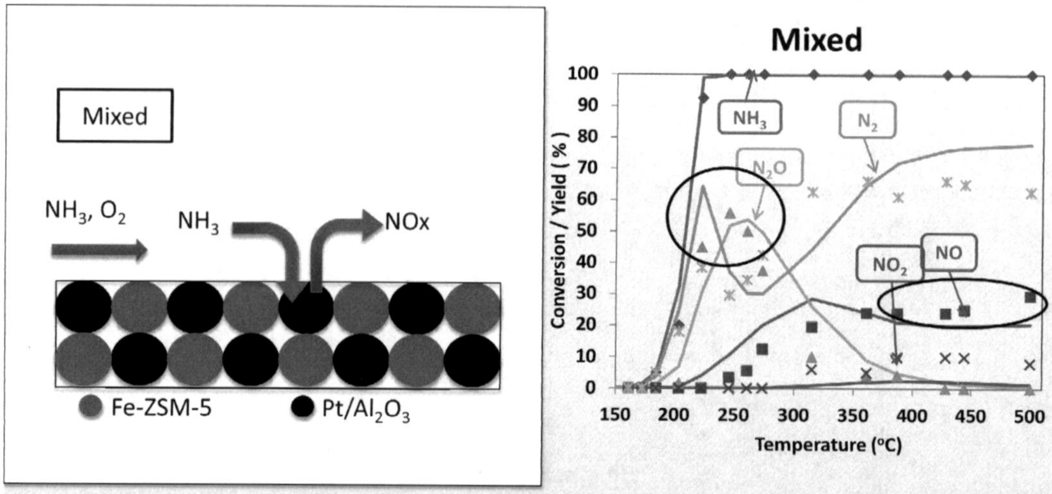

200 ppm NH_3, 5% O_2, 2.5% H_2O, 2% CO_2, bal. Ar
Experimental: Symbols
Model: Lines
GSHV 66k h^{-1}

37

Conclusions

- Lean NOx catalysts inherently multi-functional
- Layering of two catalyst functions can enhance conversion and/or selectivity
- Focused experiments combined with modeling provides proof of concept, understanding, ideas for improvement

38

Multi-scale Dynamic Simulation and Optimization for Cost-effective Industrial Emission Reductions

Xu, Q., Lamar University

Flaring emissions during CPI (chemical process industry) plant startups, shutdowns, and upsets causes tremendous raw material and energy losses and also results in adverse environmental and social impacts. For instance, an ethylene plant with 1.2 billion pounds of ethylene production per year may flare five million pounds of ethylene during one typical startup, which generates 15.4 million pounds of CO_2, 7.5 K lbs of NOX, 40 K lbs of CO, and 100 K lbs of highly reactive VOCs. Unfortunately, current practices in flare minimization almost exclusively depend on industrial experience and the "end-of-the-pipe" control strategies. For instance, the installation of flare gas recovery units (FGRU) can capture flare gas for recycle and reuse. Conceivably, it is less desirable since the "waste gas" is already generated. Meanwhile, the capital expenditure and operating cost of FGRU is considerable. Therefore, the key issue for existing CPI facilities to achieve significant emission reductions without large expenditures is to renovate current process designs and/or improve process operational strategies in a systematic way, so as to proactively and economically reduces emission sources instead of traditional "end-of-the-pipe" flare handling.

In this presentation, a systematic methodology on multi-scale dynamic simulation and optimization for cost-effective industrial emission reductions will be introduced. Since off-specification streams are inevitable during the plant turnaround operations, to significantly reduce flaring emission, they must be either recycled to the upstream process for online reuse, or stored somewhere temporarily for future reprocessing, when the plant manufacturing returns to stable operation. Thus, the off-spec products will be able to be reused instead of being flared. This presentation will address modeling and optimization for emission reduction under chemical plant start-up, shutdown, and upset conditions, as well as associated local air-quality impact studies. Both theoretical development and real plant test results could be presented.

EFFECT OF ELECTRIC FIELDS ON MAXIMUM AND MINIMUM BOILING HEAT FLUX LEVELS

Subra Sankaran

June 3, 2015

CHEMISTRY THAT MATTERS™

OUTLINE

Boiling Curve, q_minimum, Fluid Velocity, and Subcooling

Electric Field Effects

Destabilization of Film Boiling Using Electric Field - Video

Electric Fields, Boiling, and Gravity

No. 1

THE CHARACTERISTIC BOILING CURVE

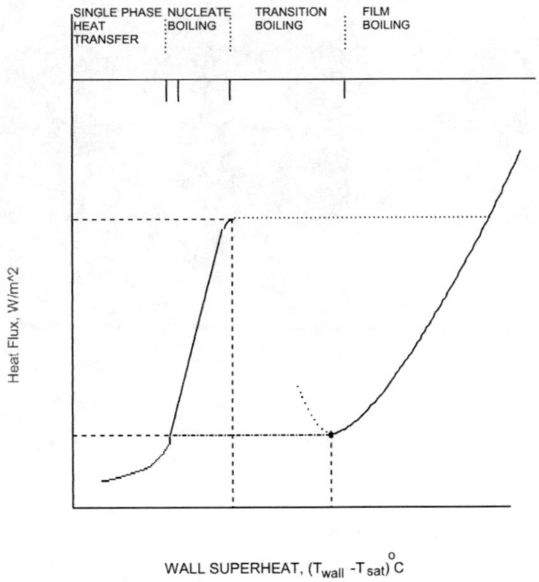

EFFECT OF VELOCITY ON Q_MAX AND Q_MIN

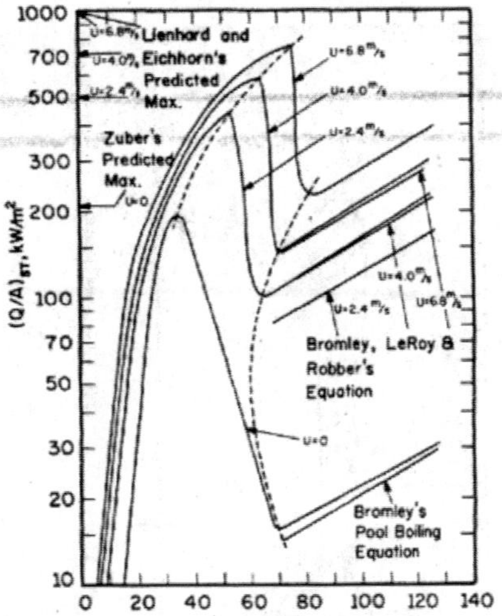

Yilmaz and Westwater – Effect of Velocity on Heat Transfer to Boiling Freon-113, J. Heat Tr., 1980

A SUBCOOLED FLOW BOILING EXPERIMENT

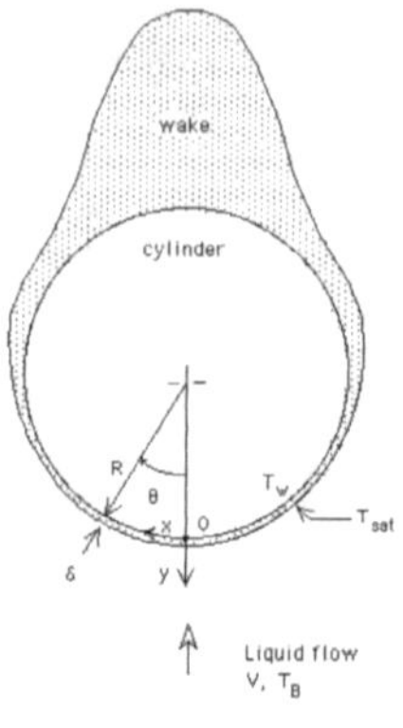

Sankaran and Witte -- 1990

EFFECT OF VELOCITY AND SUBCOOLING ON Q_MIN

Experimental and the uniform heat flux model q_minvalues.

T_sat C	Velocity m/s	T_Bulk C	ΔT_sub C	q_min_exp MW/m^2	q_min_theory MW/m^2	q_min_exp/ q_min_theory
60.0	2.95	4.4	55.6	0.772	0.628	1.15
61.5	2.95	20.6	40.9	0.515	0.458	1.12
59.5	2.95	10.6	48.9	0.626	0.551	1.14
65.5	3.53	10.3	55.2	0.878	0.679	1.29
66.0	3.53	10.0	56.0	0.886	0.689	1.29
68.9	3.81	14.4	54.5	0.904	0.694	1.30

Sankaran and Witte, 1990 -- The q_min values are 15 to 30 percent higher than the theoretical predictions. Film boiling model (Witte and Orozco, 1984), modified to constant heat flux wall (Sankaran and Witte, 1990)

3-REGION BOILING

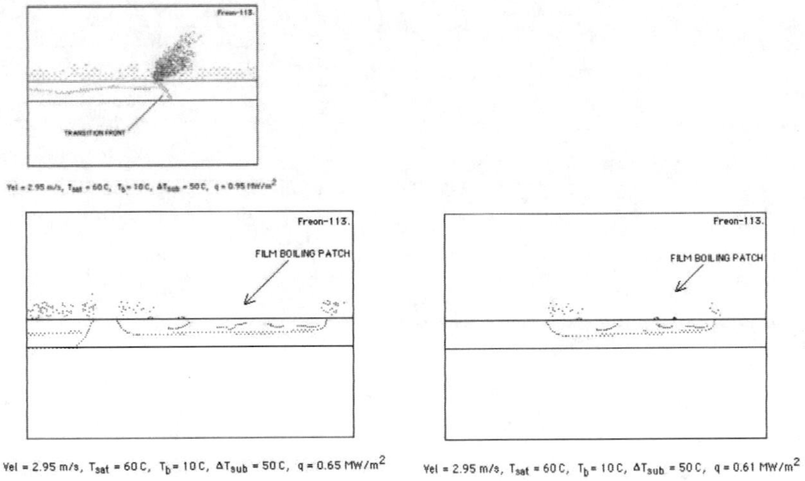

Film destabilization at the lower stagnation point in subcooled flow boiling
-- Sankaran and Witte, 1990.

No. 6

Q_MIN/Q_MAX

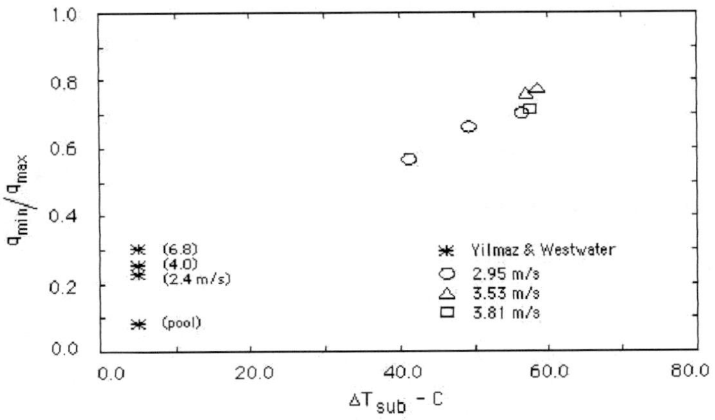

The influence of subcooling and velocity on q_min/q_max ratio.

No. 7

ELECTRIC STRESS

THE ELECTRIC STRESS TENSOR:

MAXWELL'S EQNS. FOR ELECTROSTATICS:

$$\left.\begin{aligned} \nabla\cdot\epsilon_0\epsilon E &= \rho \\ \nabla\times E &= 0 \end{aligned}\right\}$$

AT AN INTERFACE:

$$\left\{\begin{aligned} (\epsilon_0\epsilon_{OUT}E_{n_{OUT}} - \epsilon_0\epsilon_{IN}E_{n_{IN}}) &= \rho_{SURFACE} \\ E_{T_{OUT}} &= E_{T_{IN}} \end{aligned}\right\}.$$

$$\text{FORCE} = E\cdot q = \int E\cdot(\nabla\cdot\epsilon_0\epsilon E)\,dV.$$

$$\text{STRESS} = T_{ij} = \epsilon E_i E_j - \tfrac{1}{2}\epsilon E_k E_k \delta_{ij}.$$

$\therefore$

$$T_{ij} = \epsilon_0\epsilon E_i E_j - \tfrac{1}{2}\epsilon_0\epsilon E_k E_k \delta_{ij}$$

ϵ_0 = PERMITTIVITY OF FREE SPACE $= 8.854\times10^{-12}\ \frac{\text{Coulomb}^2}{N\cdot m^2}$.

ϵ = RELATIVE PERMITTIVITY OF THE MEDIUM.

$E = V/m$.

$T_{ij} = (\frac{V}{m})^2\cdot\frac{Coul^2}{N\cdot m^2} = N/m^2$.

$$F_e = \rho_e E - \frac{1}{2}E^2\nabla\varepsilon + \frac{1}{2}\nabla\left[\rho(\partial\varepsilon/\partial\rho)_T E^2\right]$$

E-FIELDS AND FILM BOILING

Table 2 Summary of theoretical results for the Cylindrical Heated Wire Problem ($b << R << R_B$, $\rho_l >> \rho_v$) ($g^* = g + \gamma/2\rho_l R^2$)

PHYSICAL MODEL	ELECTROHYDRODYNAMIC SURFACE TRACTION (τ_E)	ELECTROHYDRODYNAMIC INFLUENCE NUMBER (Ei)	DISPERSION RELATION (NORMALIZED)	NORMALIZED MOST DANGEROUS WAVELENGTH ($\underline{\lambda}_d = \lambda_d/\lambda_{d_0}$)	GROWTH RATE OF MOST DANGEROUS WAVELENGTH ($\underline{n}_d = n_d/n_{d_0}$)
(i) INSULATING FLUID, AC FIELD	$\frac{\epsilon(\frac{\epsilon}{\epsilon_0}-1)[1+(\frac{\epsilon}{\epsilon_0}-1)kR]V^2\zeta}{4R^3[\ln(\frac{R_B}{R})]^2}$	$\frac{3\epsilon(\frac{\epsilon}{\epsilon_0}-1)V^2}{8\rho_g g^* R^3[\ln(\frac{R_B}{R})]^2}$	$\underline{s}^2 + \frac{1}{2}\underline{k}^3 - \frac{1}{2}\underline{k} - Ei\underline{k} - (\frac{\epsilon}{\epsilon_0}-1)k_{d_0}REi\underline{k}^2$	$\{\frac{2}{3}Ei' + [(\frac{2}{3}Ei')^2 + \frac{2}{3}Ei + 1]^{1/2}\}^{-1}$ *	$\{\underline{\lambda}_d^{-2} + \frac{2}{3}\underline{\lambda}_d^{-1}Ei + \frac{1}{3}\underline{\lambda}_d^{-2}Ei'\}^{1/2}$ *
(ii) CONDUCTING FLUID, AC/DC	$\frac{\epsilon_0 V^2\zeta}{4b^3}$	$\frac{3\epsilon_0 V^2}{8\rho_g g^* b^3}$	$\underline{s}^2 + \frac{1}{2}\underline{k}^3 - (\frac{1}{2}+Ei)\underline{k}$	$\{1+\frac{2}{3}Ei\}^{-1/2}$	$\{1+\frac{2}{3}Ei\}^{3/4}$

* $Ei' = (\frac{\epsilon}{\epsilon_0}-1)k_{d_0}REi$

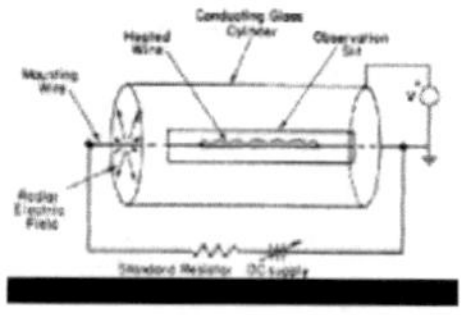

[redacted]

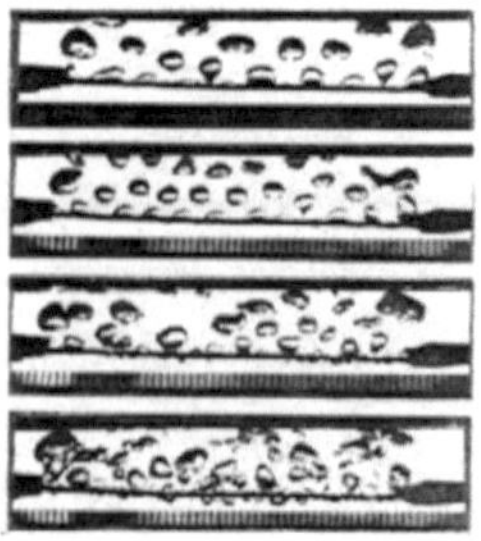

Jones and Schaeffer, 1976. Electrohydrodynamically coupled minimum film boiling in dielectric liquids. (perfect dielectric approaches)

سابك
sabic

FILM BOILING WAVE LENGTH

Rayleigh-Taylor Instability:
Most dominant wave length

$$\lambda_d \sqrt{\frac{g(\rho_f - \rho_g)}{\sigma}} = \begin{cases} 2\pi\sqrt{3} & \text{for one-dimensional waves} \\ 2\pi\sqrt{6} & \text{for two-dimensional waves} \end{cases}$$

q_min: (Zuber, Berenson – 1959)

Lienhard and Wong accounted for the transverse curvature (1964)

$$\lambda_{d,cyl} = \frac{2\pi\sqrt{3}l_L}{\sqrt{1+2/R'}}$$

Sakurai, et al. (1988)

$$\lambda_{d,cyl} = 2\pi\sqrt{3}l_L\left(\frac{2R'}{2R'+0.85}\right)$$

DiMarco (2002) et al; (a= estimated film thickness).

$$R'_{corr} = \frac{R+a}{l_L}$$

Cipriani, DiMarco, Grassi (2010)

$$\lambda_{dE} = 2\pi\sqrt{3}l_L\frac{\sqrt{3}}{El^* + \sqrt{El^{*2}+3}}$$

$$El^* = \frac{\varepsilon_{eq}\cdot E_{i,v}^2 \cdot l_L}{\sigma} = \frac{\varepsilon_{eq}\cdot E_{i,v}^2}{\sqrt{(\rho_l - \rho_v)\sigma g}}$$

No. 10

سابك
sabic

EFFECT OF GRAVITY ON FILM BOILING WAVELENGTH

436 ANNALS NEW YORK ACADEMY OF SCIENCES

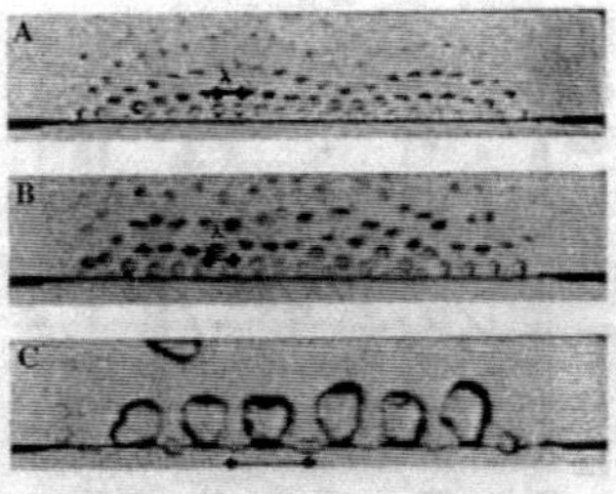

FIGURE 5. Effect of acceleration on wavelength in the absence of electric fields. Wall heat flux q = 190kW/m^2. (A) g/g_0 = 1.8. (B) g/g_0 = 1.0. (C) g/g_0 = 0.01.

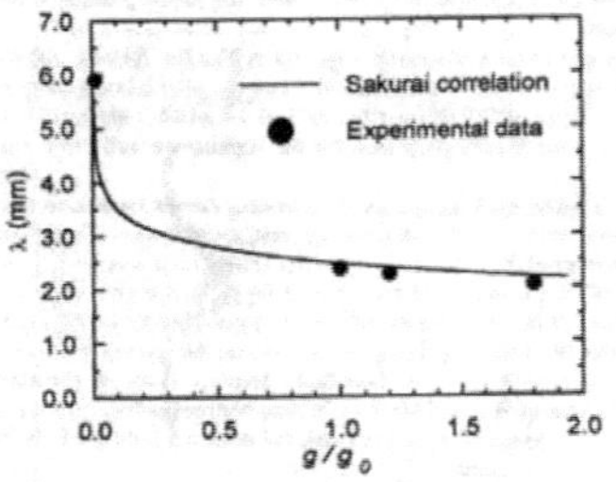

FIGURE 6. Wavelength versus gravity: comparison of the experimental data with the correlation, Equation (10), due to Sakurai *et al.*[3]

Di Marco, P, Grassi, W, Trentavizi, F., 2002,
"Pool film boiling experiments on a wire in low gravity - Preliminary results,"
Microgravity Transport Processes in Fluid, Thermal, Biological, And Materials Sciences, Ann. NY Acad. Sci., pp. 428-446.

No. 11

Q_MAX AND Q_MIN FOR VARIOUS E_FIELDS

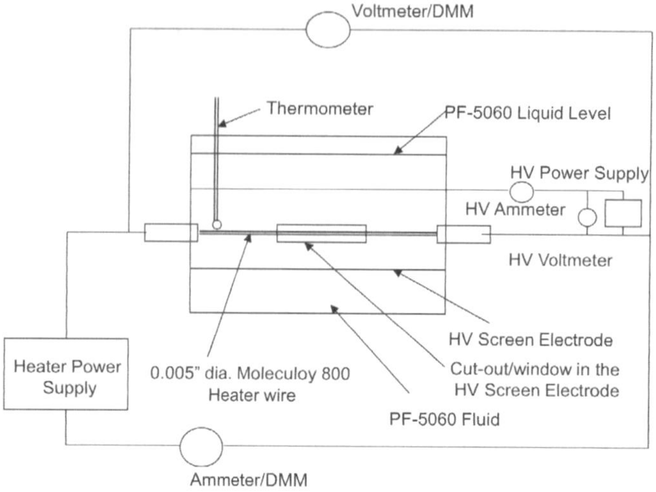

Sankaran, S.+, Allen, J.*, and Gumennik, L **– 2003 ASME Conference
+ Now at SABIC. * Now at Michigan Technological Univ., ** Now at ALCOA

No. 12

STABLE BOILING MODES TOWARDS Q_MAX

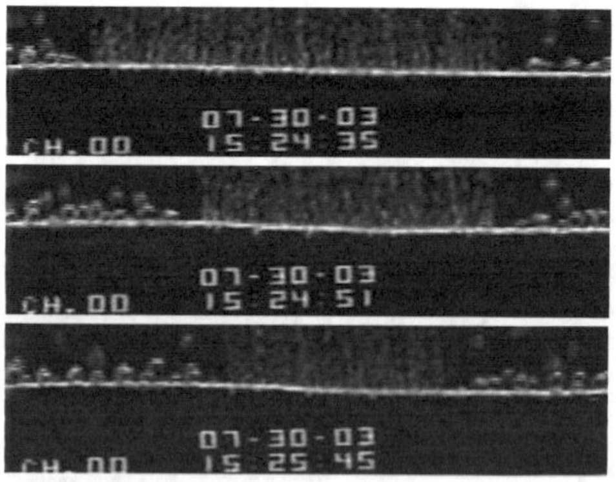

Corresponding heat fluxes:

27.6, 29.4, and 30.3 W/cm^2 respectively (T_bulk=29.4 C, T_sat=56 C, V= 1200V).
[R'=R/sqrt($\sigma/\Delta\rho g$) = 0.082] "a transition zone"

Bakhru and Lienhard ,1972 -- Observed mixed-mode boiling, and stable patchs of film boiling with water for very small radius (R' <0.01; & R'>0.15 have hydrodynamic transitions)

No. 13

FILM AND NUCLEATE BOILING

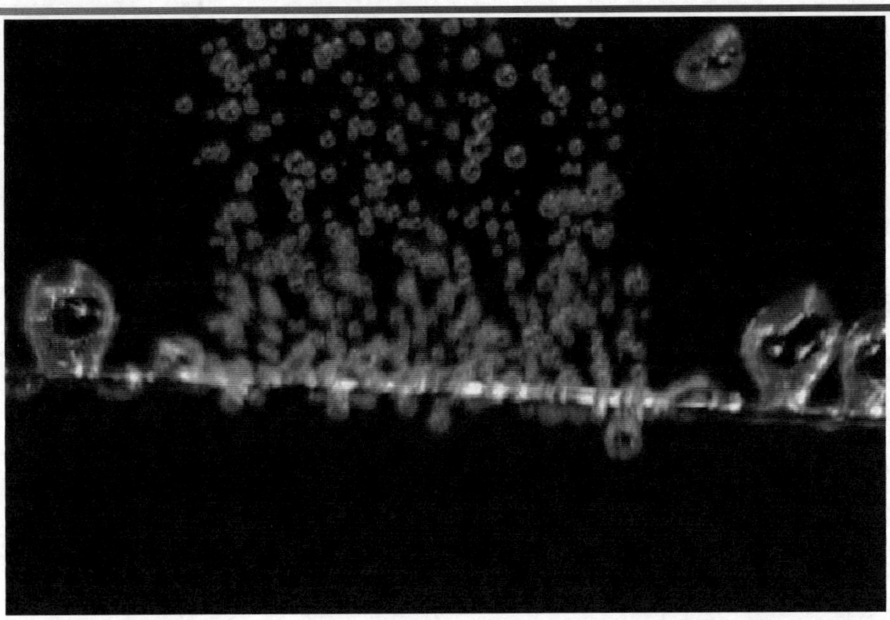

No. 14

RADIAL DC ELECTRIC FIELD DESTABILIZES FILM BOILING

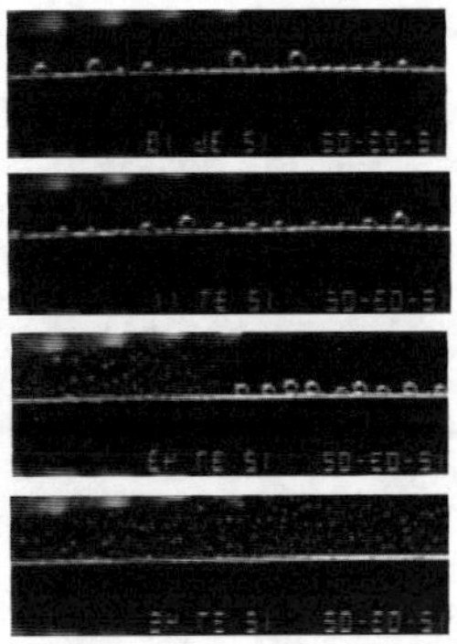

1.a
1.b
1.c
1.d

0.005” dia resistance wire in PF-5060 fluid: T_bulk ~ 80 F, q ~ 19.64 W/cm^2

1.a) Film Boiling,
1.b) film boiling with 800V between the wire and the screen
1.c) film boiling being destabilized at 1600 V,
1.d) Approximately 1 second later, film boiling completely destabilized into nucleate boiling at this 1600 V.

No. 15

EFFECT OF E-FIELD ON Q_{MIN} AND Q_{MAX}

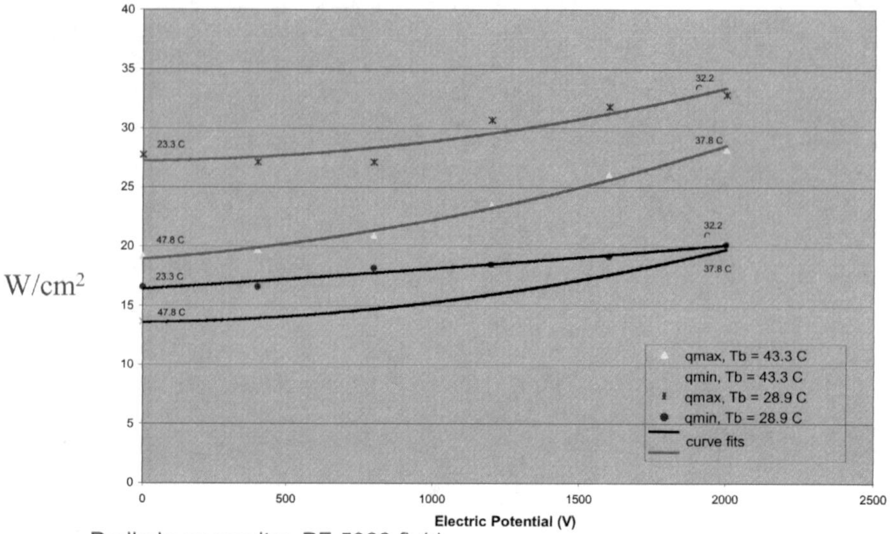

Preliminary results: PF-5060 fluid,
T_sat=56C; liquid subcooling levels are ~ 12 C and 27 C.

No. 16

SUMMARY

- q_{max} and q_{min} heat flux levels are higher by up to 50% when using 2kv across the heater and the screen, compared to the heat fluxes without electric fields.
- Use electric fields to
  - enhance heat transfer,
  - as a safety feature to actively control the onset of film boiling.
  - i.e. facilitate destabilization of vapor film by continuously driving the liquid to the heater surface.
- These will be especially beneficial for 1g, low-gravity, and microscale two-phase flow technologies.

No. 17

DISCLAIMER

DISCLAIMER: THE MATERIALS, PRODUCTS AND SERVICES OF SAUDI BASIC INDUSTRIES CORPORATION (SABIC) OR ITS SUBSIDIARIES OR AFFILIATES ("SELLER") ARE SOLD SUBJECT TO SELLER'S STANDARD CONDITIONS OF SALE, WHICH ARE AVAILABLE UPON REQUEST. INFORMATION AND RECOMMENDATIONS CONTAINED IN THIS DOCUMENT ARE GIVEN IN GOOD FAITH. HOWEVER, SELLER MAKES NO EXPRESS OR IMPLIED REPRESENTATION, WARRANTY OR GUARANTEE (i) THAT ANY RESULTS DESCRIBED IN THIS DOCUMENT WILL BE OBTAINED UNDER END-USE CONDITIONS, OR (ii) AS TO THE EFFECTIVENESS OR SAFETY OF ANY DESIGN OR APPLICATION INCORPORATING SELLER'S MATERIALS, PRODUCTS, SERVICES OR RECOMMENDATIONS. UNLESS OTHERWISE PROVIDED IN SELLER'S STANDARD CONDITIONS OF SALE, SELLER SHALL NOT BE RESPONSIBLE FOR ANY LOSS RESULTING FROM ANY USE OF ITS MATERIALS, PRODUCTS, SERVICES OR RECOMMENDATIONS DESCRIBED IN THIS DOCUMENT. Each user is responsible for making its own determination as to the suitability of Seller's materials, products, services or recommendations for the user's particular use through appropriate end-use and other testing and analysis. Nothing in any document or oral statement shall be deemed to alter or waive any provision of Seller's Standard Conditions of Sale or this Disclaimer, unless it is specifically agreed to in a writing signed by Seller. Statements by Seller concerning a possible use of any material, product, service or design do not, are not intended to, and should not be construed to grant any license under any patent or other intellectual property right of Seller or as a recommendation for the use of any material, product, service or design in a manner that infringes any patent or other intellectual property right.

† Any brands, products or services of other companies referenced in this document are the trademarks, service marks and/or trade names of their respective holders.

No. 18

PDS Process Development Symposium

AIChE
The Global Home of Chemical Engineers

4 Vital Things You Must Do to Grow Your Career

Curtis Swisher

PROCESS DEVELOPMENT SYMPOSIUM, JUNE 2-4, 2015, HOUSTON, TX

PDS Process Development Symposium

AIChE
The Global Home of Chemical Engineers

PROCESS DEVELOPMENT SYMPOSIUM, JUNE 2-4, 2015, HOUSTON, TX

AIChE
The Global Home of Chemical Engineers

Small group exercise

What are vital things you must do to grow your career?

PROCESS DEVELOPMENT SYMPOSIUM, JUNE 2-4, 2015, HOUSTON, TX

Your poll will show here

1. Install the app from pollev.com/app
2. Make sure you are in Slide Show mode

Still not working? Get help at pollev.com/app/help
or
Open poll in your web browser

AIChE
The Global Home of Chemical Engineers

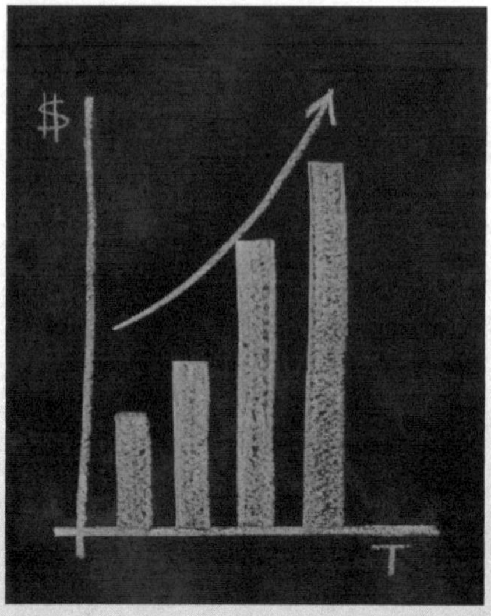

$

a/b

%

#

PROCESS DEVELOPMENT SYMPOSIUM, JUNE 2-4, 2015, HOUSTON, TX

AIChE
The Global Home of Chemical Engineers

PROCESS DEVELOPMENT SYMPOSIUM, JUNE 2-4, 2015, HOUSTON, TX

PDS Process Development Symposium

AIChE
The Global Home of Chemical Engineers

PROCESS DEVELOPMENT SYMPOSIUM, JUNE 2-4, 2015, HOUSTON, TX

PDS Process Development Symposium

AIChE
The Global Home of Chemical Engineers

PROCESS DEVELOPMENT SYMPOSIUM, JUNE 2-4, 2015, HOUSTON, TX

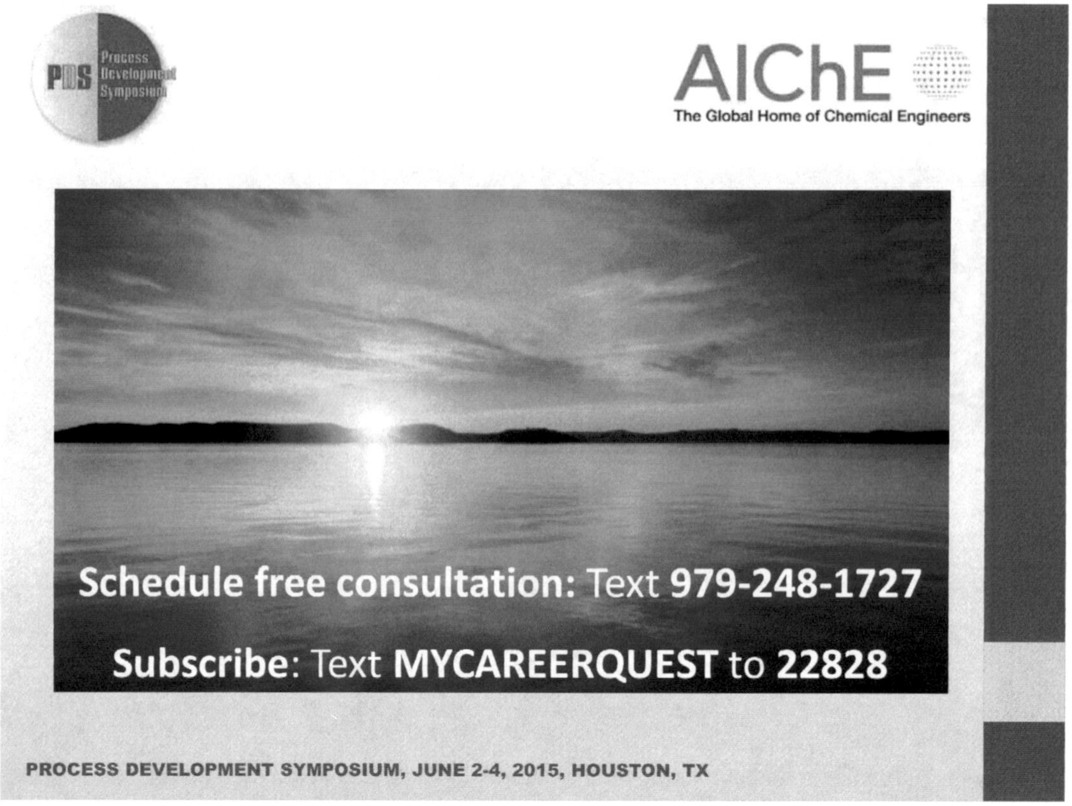
Process Development Symposium
PDS
AIChE
The Global Home of Chemical Engineers
Schedule free consultation: Text 979-248-1727
Subscribe: Text MYCAREERQUEST to 22828
PROCESS DEVELOPMENT SYMPOSIUM, JUNE 2-4, 2015, HOUSTON, TX

Countless Possibilities Of What Process Intensification Can Do

Dr. Zhijun Jia: CEO, CompRex, LLC

Prof. Dr.-Ing. Hannsjörg Freund: Friedrich-Alexander University Erlangen-Nürnberg

June 3, 2015, Houston, TX

AIChE Process Development Symposium

COMPREX

CompRex™ Compact Heat Exchange Reactor

FAU FRIEDRICH-ALEXANDER UNIVERSITÄT ERLANGEN-NÜRNBERG TECHNISCHE FAKULTÄT

Outline

- Introduction to Process Intensification
- Current Applications of Process Intensification
- Examples of Process Intensification
- Future Trends of Process Intensification
- Summary

Process Intensification: A few remarks

- Young and evolving field in chemical industry
- Rapidly increasing interest in academia and industry (esp. in Europe)
- Conferences, papers, projects, roadmaps...
- Lots of examples for improved processes
- Collecting examples and classifying them in schemes

Process Intensification: How to define?

- Some counter: PI = Process Intensification or Process Optimization
- General associations with PI:
  - Making more with less
  - Cheaper, smaller, cleaner
  - Faster (reduce time-to-market)
  - Significant/drastic improvement
- These are the benefits of PI, but how to apply???

Process Intensification: What is the essence?

- PI requires a holistic view of a process, with aim at significant process improvements
- Looking at each process at a more fundamental level
- Thinking in terms of functions rather than unit operations
- Considering the whole spectrum of physico-chemical phenomena
- Adopt new materials, tools and equipment from other disciplines
- Challenging but important: Systematic approach for classification and comparison of PI options

see: H. Freund, K. Sundmacher, *Chem. Eng. Process.*, 2008, 2051-2060

Process Intensification: Examples

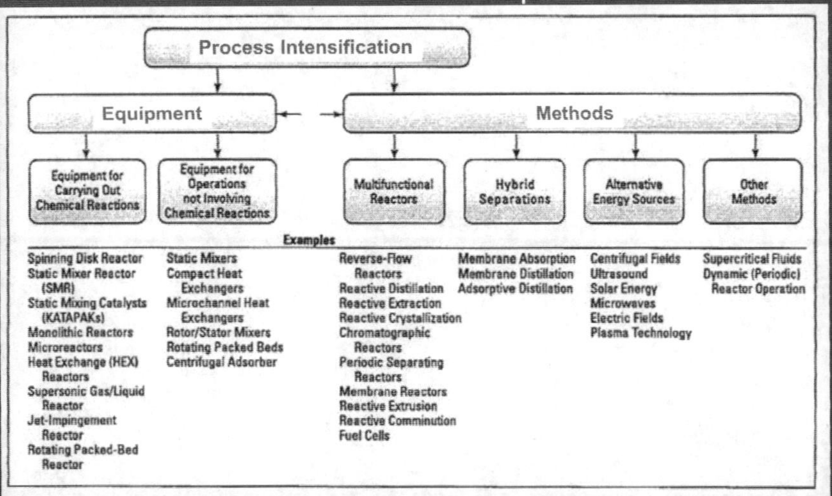

see: A.I. Stankiewicz, J.A. Moulijn, *Chem. Eng. Progress*, 2000, 22-34

Process Intensification: Process Hierarchy

- The four levels of PI
  - Plant level (overall process flowsheet)
    - Integration concepts: Heat integration, reactive separations
  - Process unit level (optimal process route)
    - Reactor design (SA:V, Δp), catalyst geometry/shape
  - Phase level
    - Thermodynamic – Tuning of phase behavior (solubility, miscibility)
    - Kinetic – Tuning of kinetic coefficients for reaction, mass & heat transfer
  - Molecular level
    - Reaction route design
    - Catalyst design

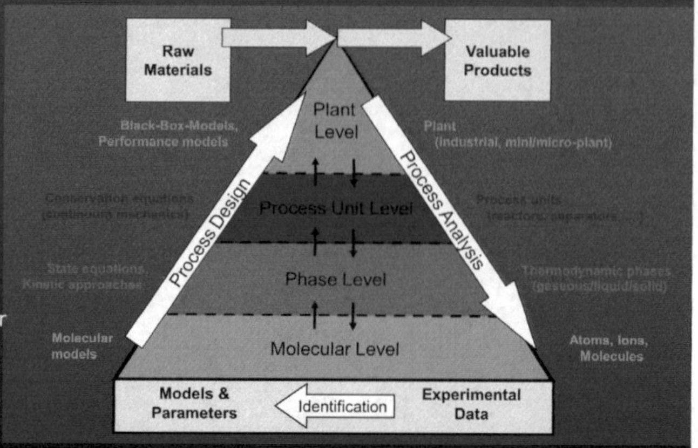

see: H. Freund, K. Sundmacher: Process Intensification I-IV, *Ullmann's Encyclopedia of Industrial Chemistry*, Wiley-VCH, 2011

Example of PI Strategies

- Reduce plant size
- Reduce unit operation sizes and/or number of unit ops
- Integrate unit operations into multi-functional units
- Make use of modular (standardized) units & processes
- From (large) batch equipment to (smaller) continuous units (esp. pharmaceuticals & fine chemicals)

Advantages of PI Implementation

- Reduce CapEx and OpEx
- Reduce energy usage
- Reduce chemical inventory
- Reduce waste streams
- Reduce time-to-market
- Improve plant safety and simplify operations

Ex: Eastman Kodak Process – Methyl Acetate Synthesis

From 1 reactor + 10 separation units to <u>one single **integrated** column</u>

MeOH + HAc <=> MeOAc + H_2O

Katalysator: H_2SO_4/ saurer Ionenaustauscher
$K_x(25°C) = 5.2$
Siedetemperatur bei p=0.1 M Pa T_s:

Azeotrop MeAc / MeOH	53.8 °C
Azeotrop MeAc / H_2O	56.7 °C
MeAc	56.9 °C
MeOH	64.6 °C
H_2O	100.0 °C
HAc	118.0 °C

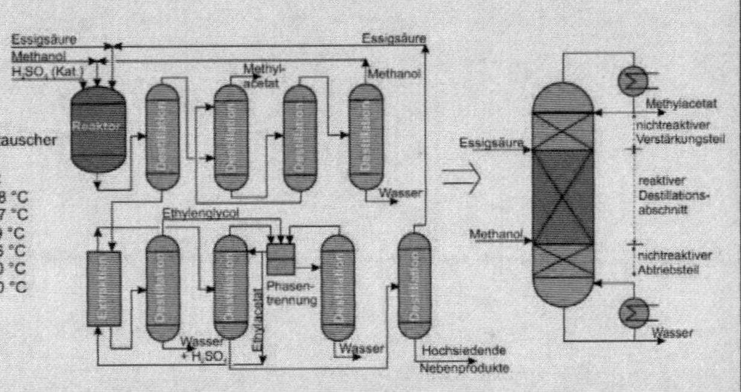

see e.g. Sundmacher, K., Ivanova, M., *Chemie in unserer Zeit* 37 (2003) 268 – 278

Ex: BASIL Process – Biphasic Acid Scavenging utilizing Ionic Liquids (IL)

Traditional Synthesis of Alkoxyphenylphosphines

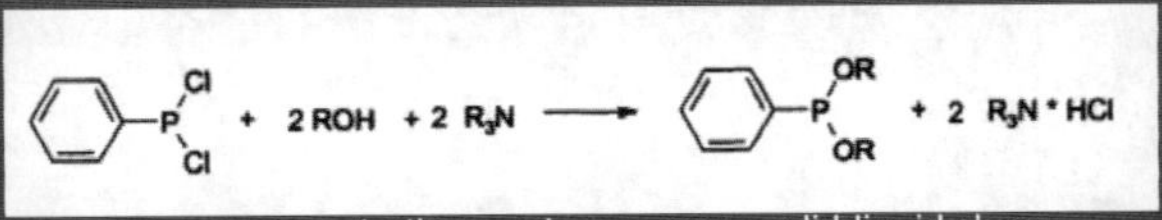

tertiary amine

solid-liquid slurry

Acid Scavenger

STY increase by factor 8 x 10^4

Synthesis of Alkoxyphenylphosphines with IL

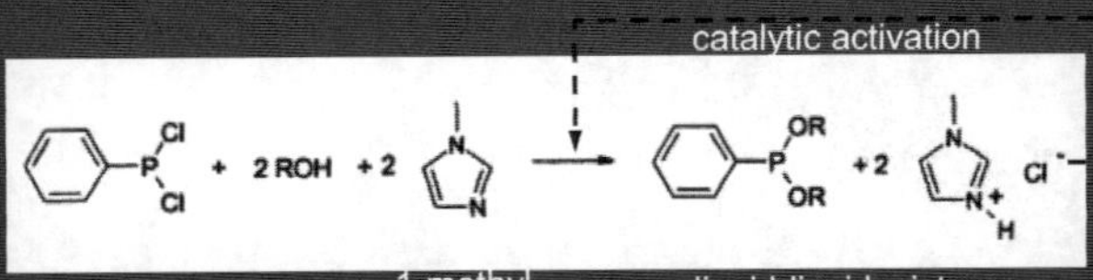

1-methyl imidazole (IL)

liquid-liquid mixture

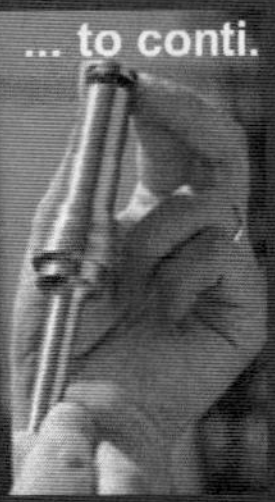

see e.g.: Masse, M., Massonne, K., *ACS Symp. Series* 902 (2005) 126-132

Ex: Compact Heat Exchanger Design

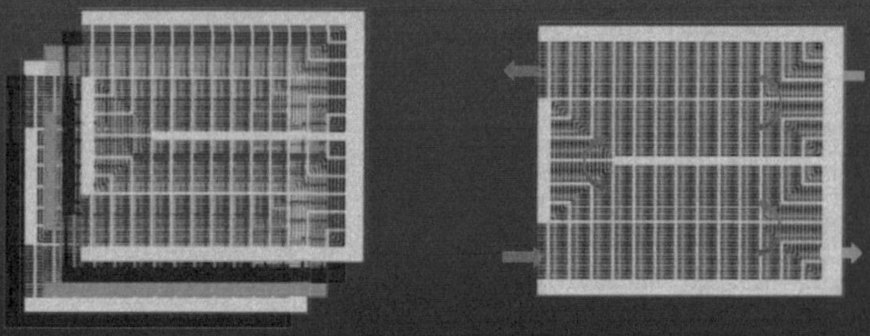

3D Dolphin Flow for improved heat transfer

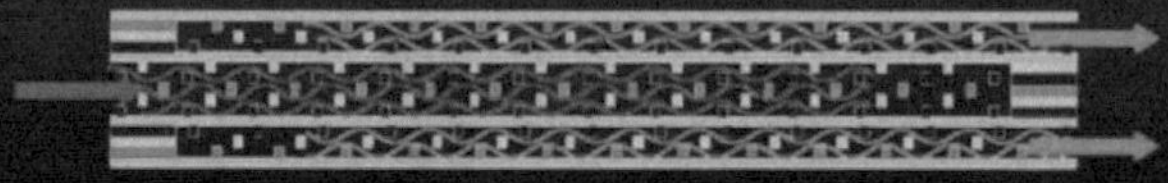

6/3/2015 AIChE Process Development Symposium: Process Intensification 11

Ex: Application in Brayton Cycle using sCO_2

- Supercritical CO_2
  - Fluid properties near and above critical point more advantageous than H_2O in Rankine Cycle
- Higher T difference for more efficient heat transfer
  - Compact heat exchangers
  - Smaller footprint
- Higher energy density than water
  - Less fluid flow
  - Smaller footprint

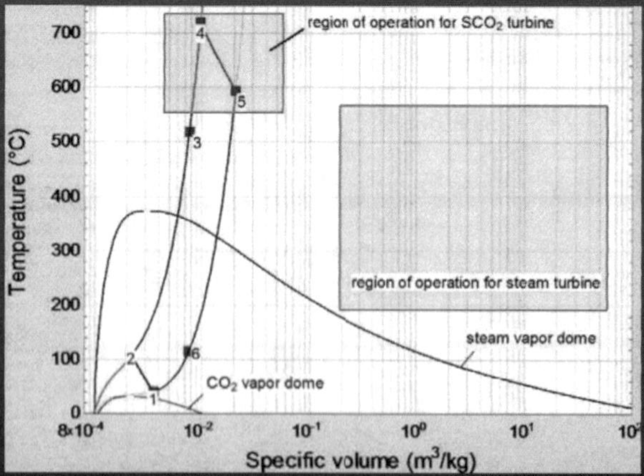

6/3/2015 AIChE Process Development Symposium: Process Intensification 12

Ex: Brayton Cycle with Compact Heat Exchanger

- Compact equipment
  - Compact heat exchangers
  - Compact sCO_2 turbines
- Smaller overall plant footprint
- With higher allowable T (material limited) efficiencies can well exceed H_2O Rankine Cycle
- Efficiency increase – reduction of fuel input, increase energy output
- Smaller & lighter equipment for mobile energy/power production
  - Marine application
  - Aerospace application

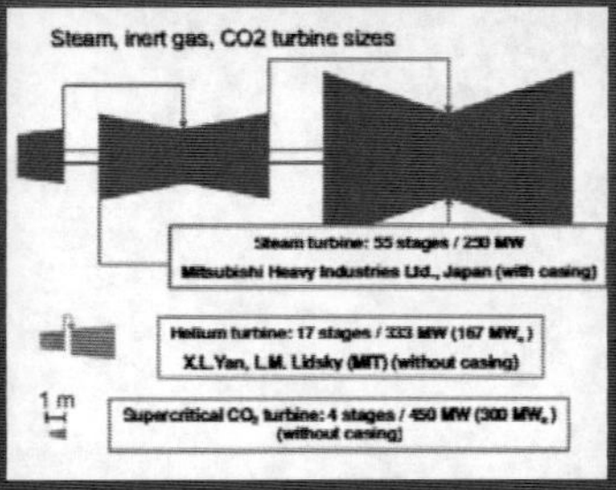

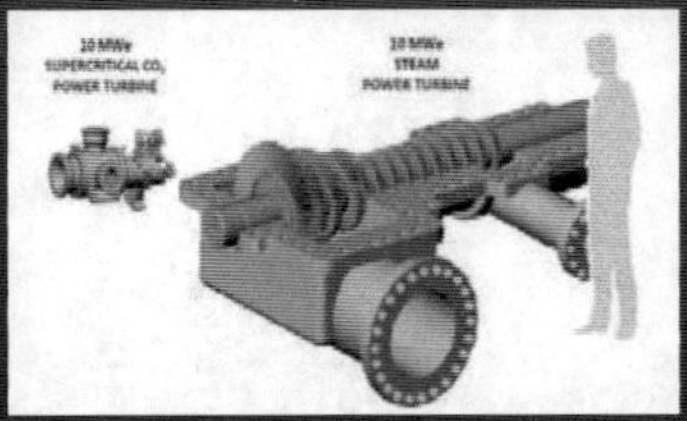

Ex: Compact Heat Exchanger Reactor Design

- Compact heat transfer integration
  - Brazed Plate-fin exchanger reactor
  - Diffusion bonded exchanger reactor
- Intermittent injection: Adding chemical at strategic locations

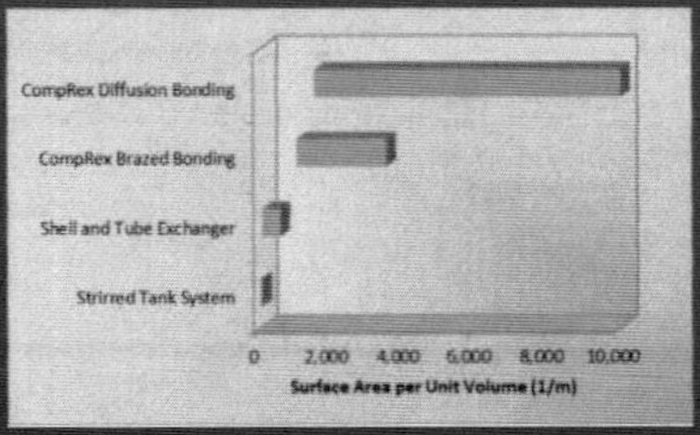

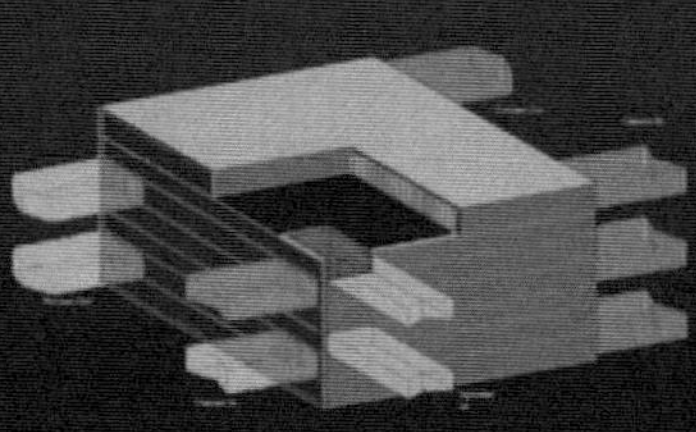

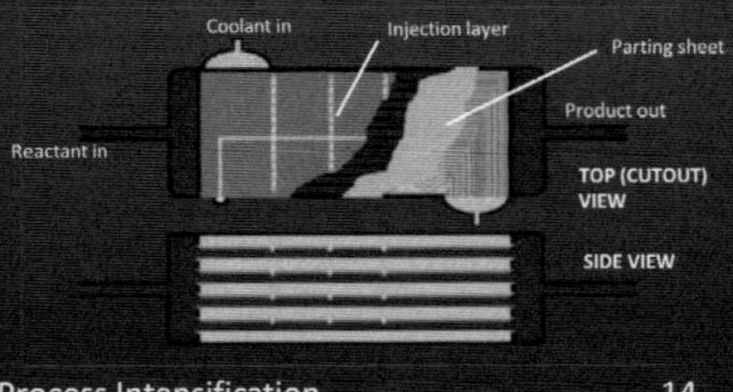

Ex: Applications for Compact Heat Exchanger Reactor

Chemical reactions where tight temperature control is critical for process, such as

- XTL – Fischer Tropsch
- Alcohol production
- Polymerization
- Oxidation
- Hydrogenation
- Hydrogen production

Ex: Conventional FT Reactors

Slurry Bed (Ex. Sasol Oryx)

- Good heat transfer
- Potential mass transfer limitation
- Difficult removing larger viscous products (i.e. wax) from catalyst

- 33ft diameter, 100ft long produces 17,500bpd syn-crude

Fixed Bed (Ex. Shell Pearl)

- Simple to operate
- Moderate ease of scale up
- Poor temperature control

- 2,800 1in tube reactors 40 ft long produces 7,500bpd syn-crude

EX: Compact FT Reactor

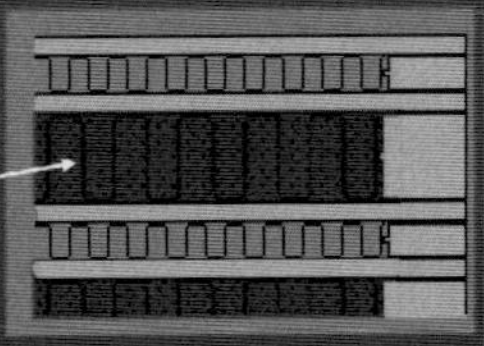

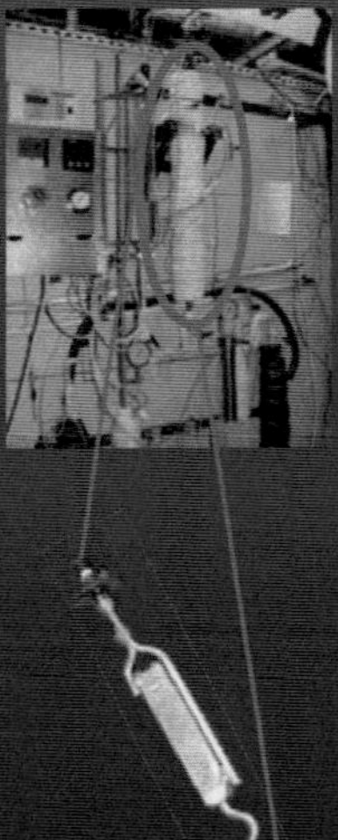

4 feet long reactor can reach the conversion same as in 40 feet long conventional tubular reactor

Comparison of compact reactor and CSTR

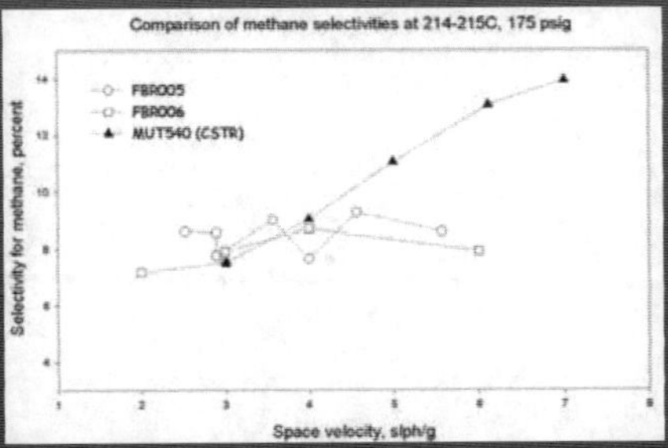

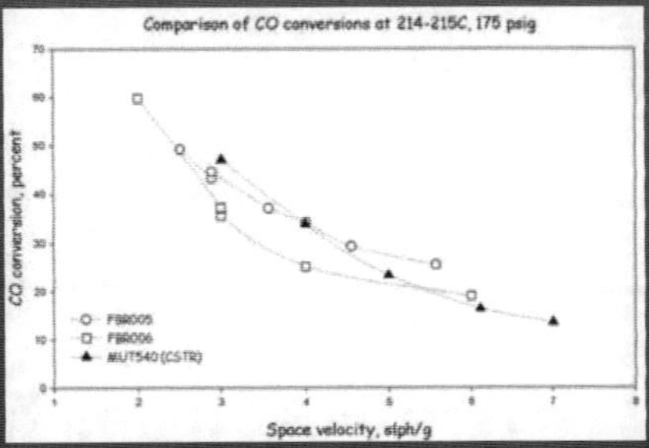

Similar conversion in compact heat exchanger reactor and CSTR indicates:

- No heat transfer limitations observed in compact reactor
- Meet the maximum productivity for the catalyst

Ex: Alcohol Production in Compact SS Reactor

Bench scale test facility:

Production rate per unit of catalyst is 4~5 times higher than the conventional tubular reactor

50-GPD Synthesis Facility

6/3/2015 AIChE Process Development Symposium: Process Intensification 19

Ex: Alcohol Production in Compact SS reactor

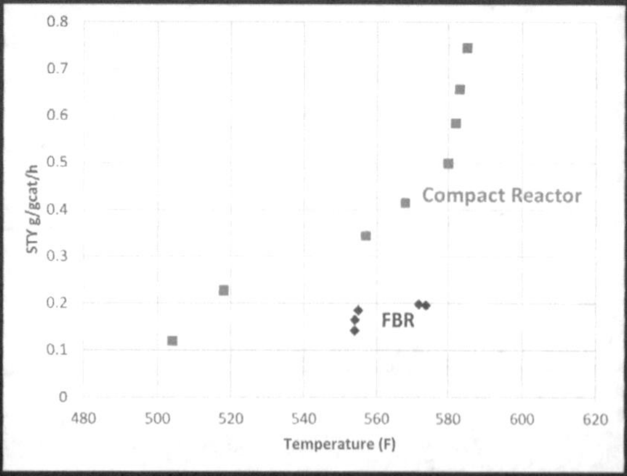

- Catalyst utilization:
  5X intensification over FBR
  - Highest Productivity/Reactor Volume
- Improved conversion observed over 2-year operation

6/3/2015 AIChE Process Development Symposium: Process Intensification 20

Countless Opportunities for PI……

Considering at least four levels in PI applications:

- Molecular --- Fundamental to lab scale
- Phase --- Lab to bench scale
- Process Unit --- Bench to pilot scale
- Plant --- Pilot to industrial scale

PI Options at the Molecular Level[1]

Planning of Reaction Routes

- Selection of Main Reactant Molecule ‘A‘
  - Atomicity & Transformation efficiency
  - Feedstock
- Selection of Partner Reactant Molecule ‘B‘
  - Avoid couple products
  - Avoid side reactions

Selection of the Catalyst

- Dream Reactions
  - Methane, CO_2, Benzene, Renewables, ...
- Molecular Design of Hom. & Het. Catalysts
  - Quantum Chemical Calculations
  - Atomistic Calculations

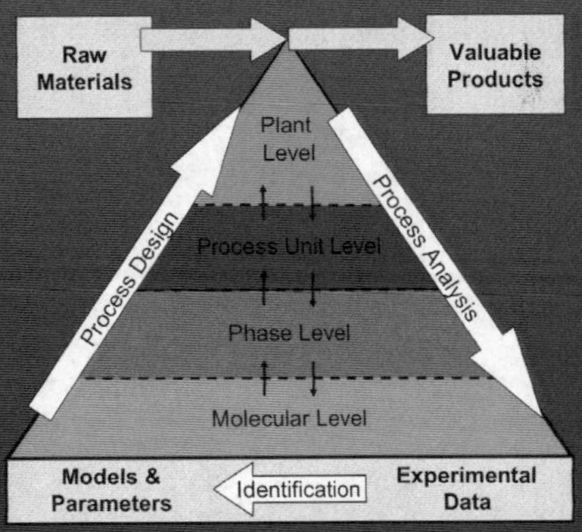

[1] H. Freund & K. Sundmacher, *Process Intensification I*, In: Ullmann's Encyclopedia of Industrial Chemistry, Wiley-VCH, 2011

PI Options at the Phase Level[1]

Thermodynamic Methods

- Property Shifting by Molecular Design
  - Ionic Liquid Systems
  - Solid-Bound Systems
- Property Shifting by p, T, and Solvent Composition
  - Supercritical Fluids
  - T-Dependent Multicomponent Solvent Systems

Kinetic Methods

- Electromagnetic Tools
  - Photochemical Induction
  - Microwave Dielectric Heating
- Fluid-Mechanical Tools
  - Cavitation
  - Centrifugal Forces

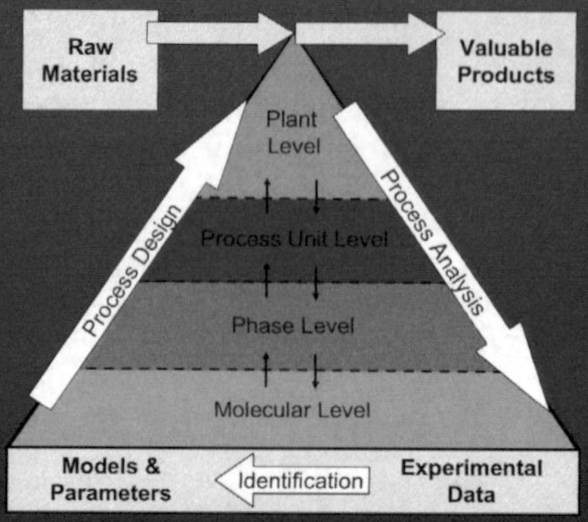

[1] H. Freund & K. Sundmacher, *Process Intensification II*, In: Ullmann's Encyclopedia of Industrial Chemistry, Wiley-VCH, 2011

PI Options at the Process Unit Level[1]

Miniaturization

- Micro & Milli vs. Macro Structured Equipment

  Functions & Equipment
  - Contacting: Micro Mixers
  - Activating: Micro Heat Exchangers
  - Reacting: Micro Reactors
  - Separating: Micro Separators

Structured Apparatuses

- Internal Structuring
  - Regular vs. Random Structures
  - Divided Wall Column
- External Structuring
  - Primary vs. Secondary Geometry Effects

Novel Flow Regimes

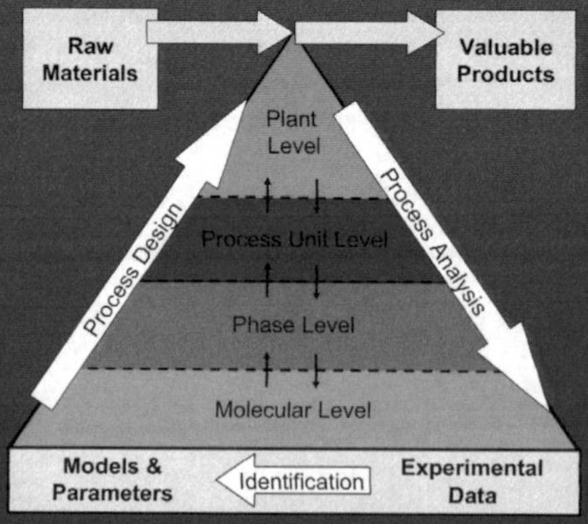

[1] H. Freund & K. Sundmacher, *Process Intensification III*, In: Ullmann's Encyclopedia of Industrial Chemistry, Wiley-VCH, 2011

PI Options at the Plant Level[1]

Heat Integrated Chemical Reactors
- Combination of Exo- & Endothermic Rxns.
- Desorptive Cooling

Reactive Separations
- Reactive Distillation
- Reactive Extraction
- Reactive Absorption
- Reactive Crystallization
- Membrane Reactors

Further Integration Concepts
- Hybrid Separations
- Reactive Mechanical Unit Operations
- System Integration

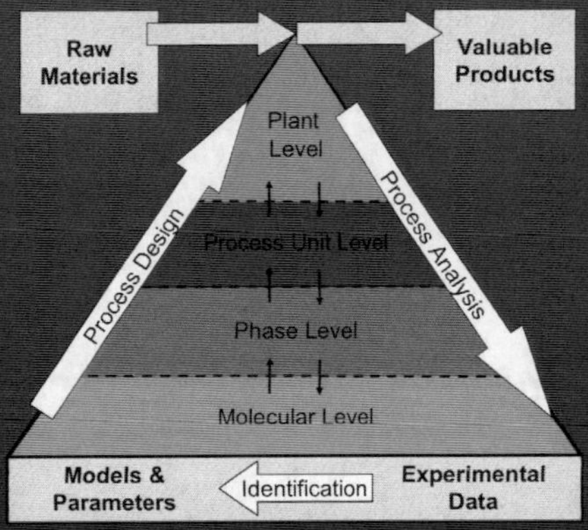

[1] H. Freund & K. Sundmacher, *Process Intensification IV*, In: Ullmann's Encyclopedia of Industrial Chemistry, Wiley-VCH, 2011

6/3/2015 AIChE Process Development Symposium: Process Intensification 25

Technological Hurdles

- As Process Intensification pushes through current limitations, other technological hurdles and opportunities present themselves
  - Catalyst particle size for compact channel reactors
    - Larger than catalysts in Fluidized Bed reactors
    - Smaller than conventional Fixed Bed reactors
  - Materials and alloys for higher pressure and higher temperature applications
    - Brayton cycle efficiency increases at higher temperature
      - Nuclear energy temperature source (up to 750C)
      - Solar energy temperature source (up to 1,000C)
  - More active catalysts, especially T-sensitive to selectivity?
  - Operability of highly integrated processes?
  - ...

6/3/2015 AIChE Process Development Symposium: Process Intensification 26

Summary

- Process Intensification is a young and evolving technology-modifying paradigm
- The purpose of PI is to significantly improve processes for lower costs and safer operations
- Many processes have been identified for improvement through PI
- Many more processes will be identified for PI
- For this, a systematic approach to PI is essential
- PI also offers technological opportunities to further push the intensification steps
  - Stronger materials, more active catalysts, etc...

Acknowledgement

- Test for FT reactor was completed by the group led by Burtron Davis at University of Kentucky, a partner of CompRex
- Test for alcohol product in SS reactor was completed by Vijay Sethi at Western Research Institute, a partner of CompRex

SRNL-STI-2015-00267

We put science to work.™

Savannah River National Laboratory™

OPERATED BY SAVANNAH RIVER NUCLEAR SOLUTIONS

European Efforts in Process Intensification

Bond Calloway, Thad Adams, and Kevin Fox

Savannah River NUCLEAR SOLUTIONS

Acknowledgements

- Dr. Adam Harvey, Newcastle University
- Prof. Andrzej Stankiewicz, Director, TU Delft Process Technology Institute
- Solvay Hosts, Dr. Jean-Pierre Brunelle, Vice President, Process Innovation
- DSM Hosts, Dr. Peter Jansens, Director, DSM ChemTech R&D and Corporate Scientist Process Technology

SRNL is Critical to DOE Success

- Strategic partner at other DOE Sites
- Over $5 billion in projected savings in past five years

- Fukushima support
- Constructed Wetlands, in partnership with Clemson

Mission: Treatment & Disposal of 92 Million Gallons

Waste Processing: Treatment and Disposal of Radioactive Waste
Mission: Treat 92 million gallons (343 million liters)
505 million curies of radioactive tank waste (7.39 x 10^{18} becquerels)

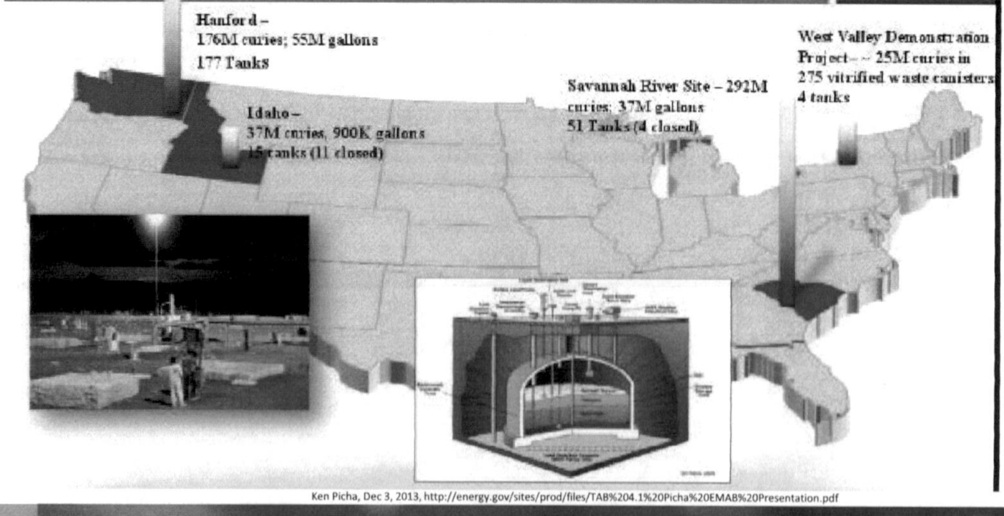

Ken Picha, Dec 3, 2013, http://energy.gov/sites/prod/files/TAB%204.1%20Picha%20EMAB%20Presentation.pdf

How Much is 92 Million Gallons?

375' Diameter, 40' Tall, 32 Million Gallons

A Large Nuclear Chemical Complex

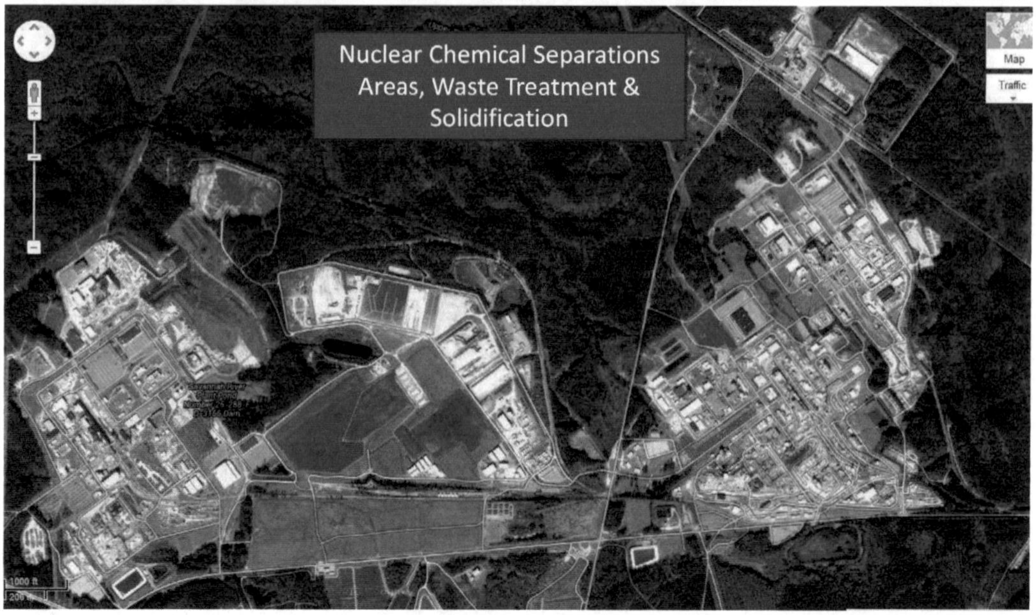

> 3 Miles

Flowsheet

Chemistry
- Inorganic
- Many trace organics
- Most all elements represented
- Solid, Liquids, Slurries(Gas-Liquid-Solid), Gases

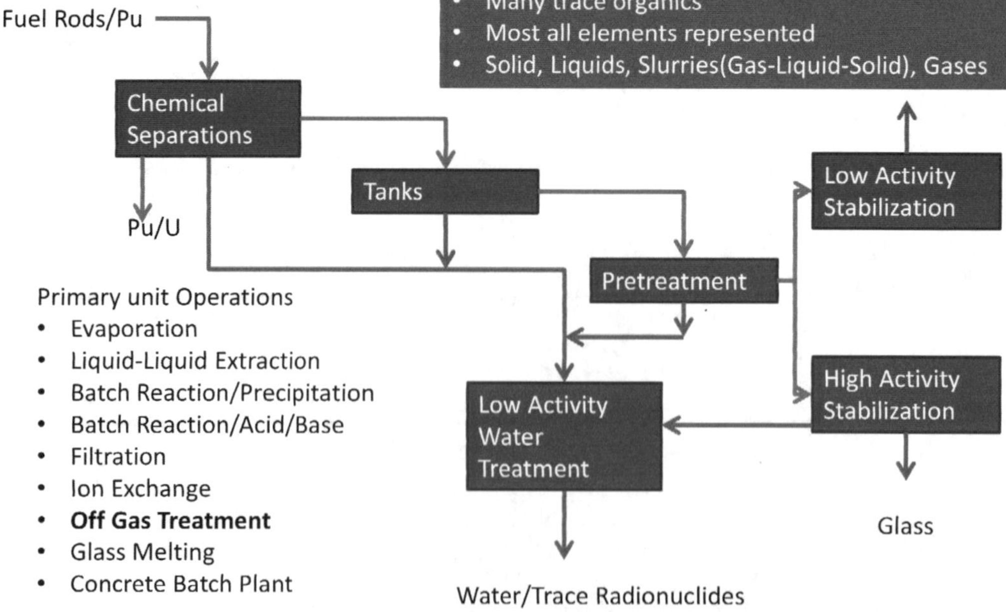

Primary unit Operations
- Evaporation
- Liquid-Liquid Extraction
- Batch Reaction/Precipitation
- Batch Reaction/Acid/Base
- Filtration
- Ion Exchange
- **Off Gas Treatment**
- Glass Melting
- Concrete Batch Plant

Canyon Cross-Section-Can We Make this Smaller?

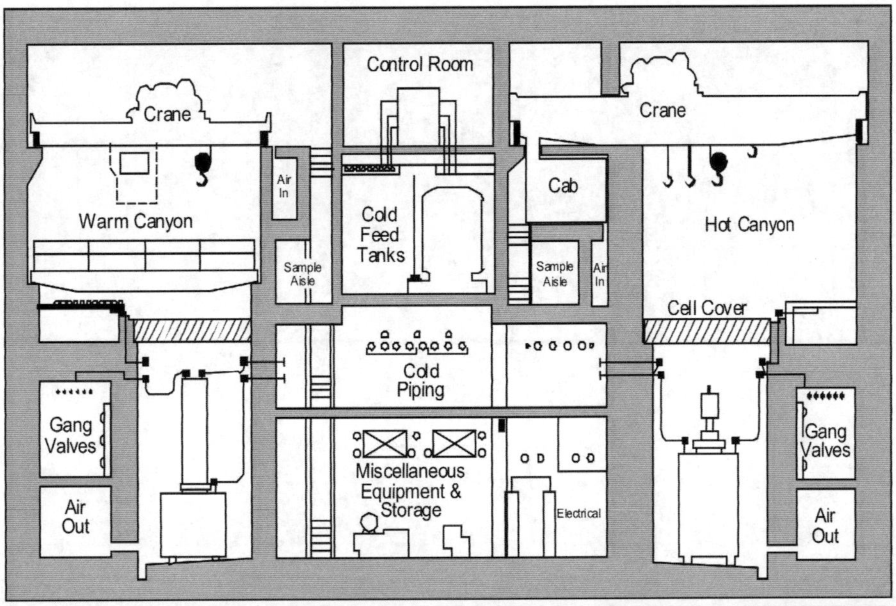

Savannah River National Laboratory
SRNL-STI-2015-00267
We put science to work.
9

Piping Gallery in Canyon Chemical Separations Facility

Savannah River National Laboratory
SRNL-STI-2015-00267
We put science to work.
10

Another Multi-Billion Dollar Site Under Construction

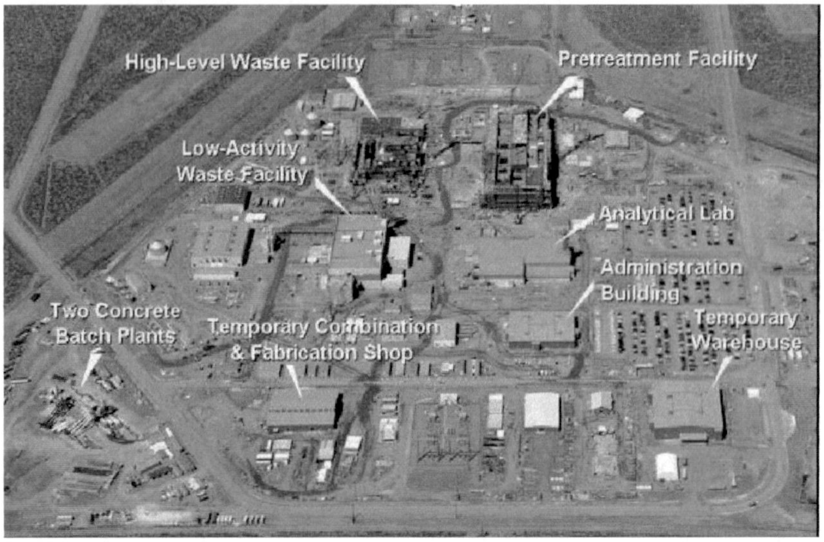

Nuclear Waste Clean-up Faces Major Challenges

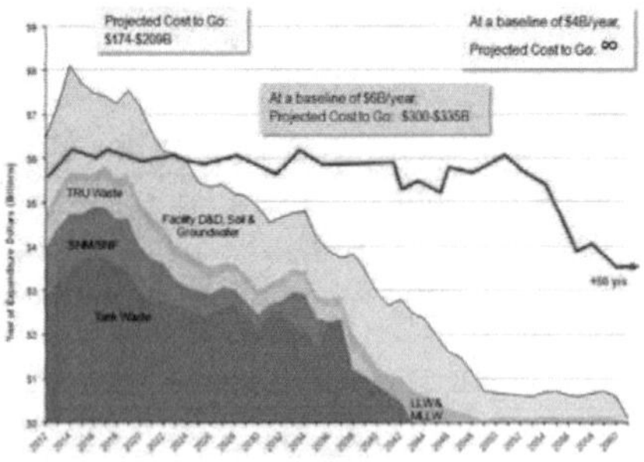

Today, EM lifecycle cost is a function of Funding level.

- Meeting baseline regulatory agreements will require unrealistic funding.
- Optimistic funding levels push cleanup schedule past 2070.
- As cleanup schedule extends, maintenance and infrastructure consume increasing fraction of available funds.

Nuclear Chemical Processing is 30+ Years Behind the Chemical Industry

Old Technology	Old Controls
Bigger is better; Cost ~ Production Capacity$^{2/3}$	Output Based Process Control; Production decisions made offline

Process Intensification: Benchmarking Effort: Objectives

- Hold introductory information exchanges with the leading European PI organizations and assess the ability to leverage the technologies developed by commercial companies for the benefit of DOE-EM/NNSA;
- Evaluate EU efforts and foster direct collaboration with EU institution, so as to benefit the DOE-EM/NNSA mission and SRNL's efforts to transform the US nuclear chemical industry through PI
- Determine the engineering methodologies used to implement PI by modern chemical companies
- Determine whether Smart Manufacturing concepts are being implemented with European chemical companies.

In Europe awareness of the importance of PI has grown strongly in recent years

- PI placed clearly in sustainability context
- First university courses, books, international conferences, journal
- National networks in UK, NL and DE
- Numerous industrial initiatives
- EFCE Working Party on Process Intensification
- European Roadmap for PI
- PI in European programs

europic

Drivers - EU Interest in PI

- Economic
- Environmental
- Sustainability
- Social Pressure (Chemical Manufacturing not unlike Nuclear Power)

In 2012

The core of the chemical industry is shifting to Asia, and by 2030 at least half of the top 10 chemical companies will be Asian or Middle Eastern

Sales

(€ billion; 2030 is calculated at 2010 prices and exchange rates)

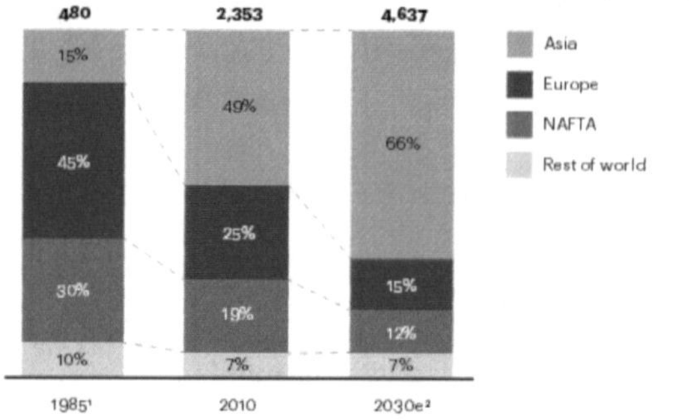

http://www.atkearney.com/documents/10192/536196/Chemical+Industry+Vision+2030+A+European+Perspective.pdf/7178b150-22d9-4b50-9125-1f1b3a9361ef

Even with Shale Gas, trend is not expected to change in Europe – Increase in US expected

Savannah River National Laboratory SRNL-STI-2015-00267 We put science to work. 15

Some European Centers of Excellence in Process Intensification

EUROPIC/Delft University of Technology	Industry Consortium /Delft, Netherlands
DSM	Chemical Company/Geleen, Netherlands
Newcastle University/Process Intensification Network	University/Newcastle, UK
Solvay	Chemical Company/Lyon, France
Bayer- Dormund University of Technology INVITE Center	Industry Center/Leverkusen, Germany
Institute for Micro Process Engineering/Karlsruhe Institute of Technology	University-National Laboratory/ Karlsruhe Germany

Savannah River National Laboratory SRNL-STI-2015-00267 We put science to work. 16

Process Intensification has been included in the Seventh Framework Program of the European Commission

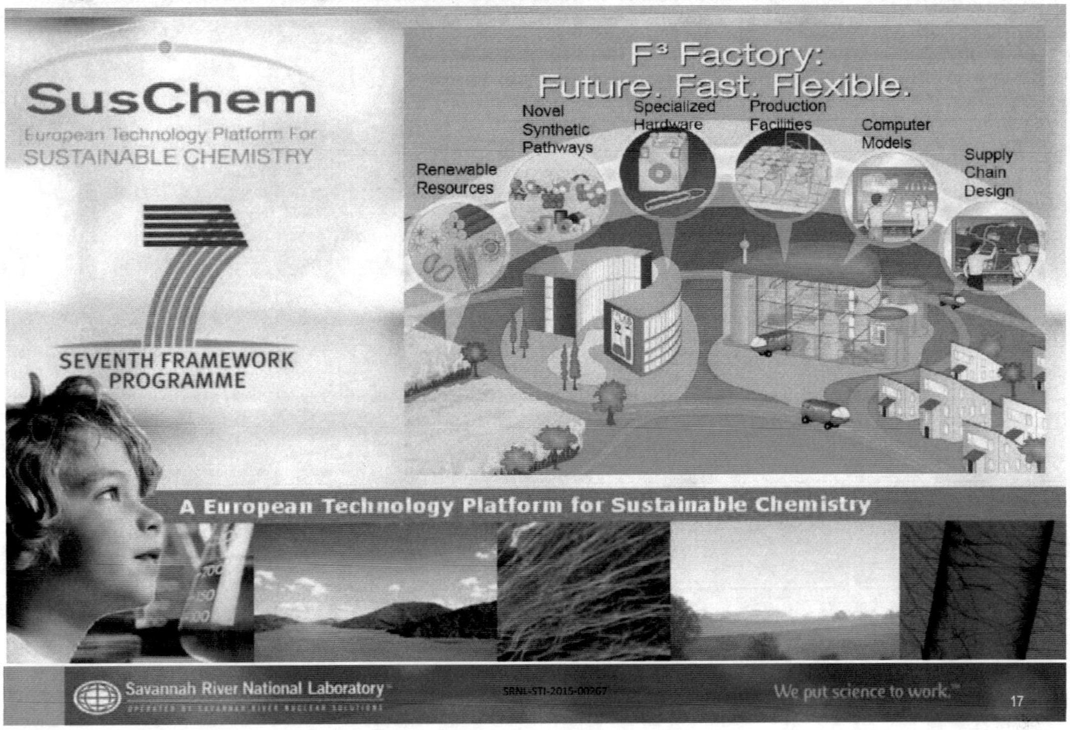

EUROPIC:
An industry-driven platform for knowledge and technology transfer in PI

Delft - Dortmund - Toulouse

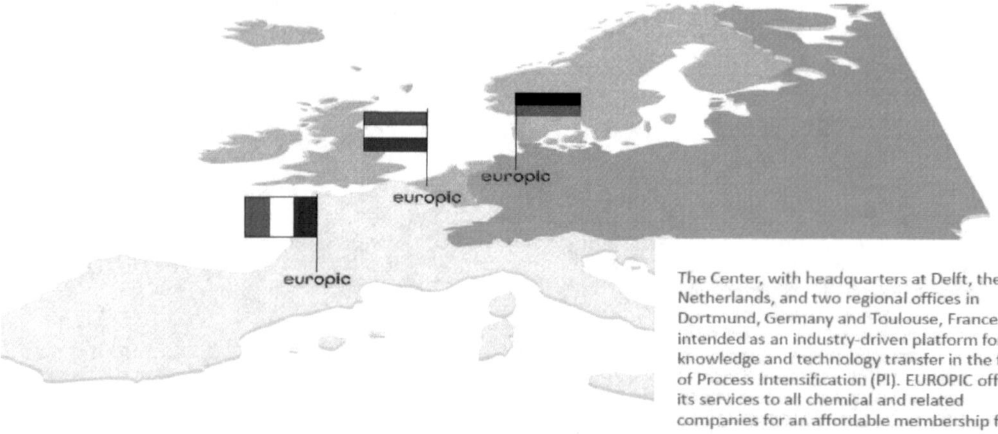

The Center, with headquarters at Delft, the Netherlands, and two regional offices in Dortmund, Germany and Toulouse, France is intended as an industry-driven platform for knowledge and technology transfer in the field of Process Intensification (PI). EUROPIC offers its services to all chemical and related companies for an affordable membership fee.

europic

Analyses of PI technologies

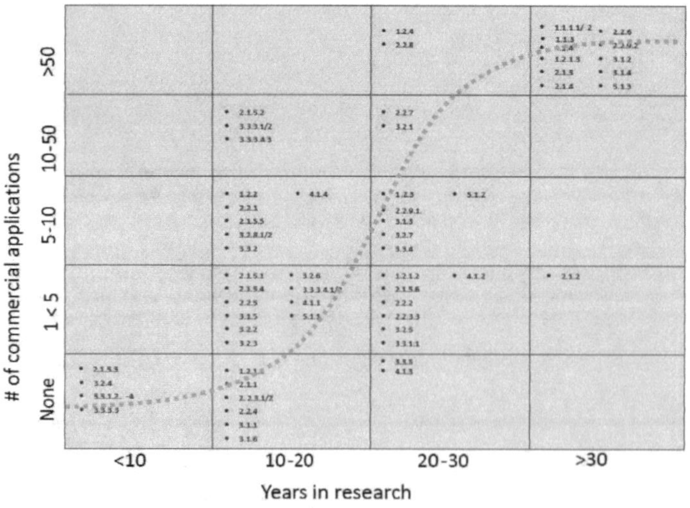

PI-technologies analyzed on number of reported commercial applications versus time in research.

Positive and negative deviations from the average technology development curve can be seen.

Number codes correspond to individual technology or equipment in our database.

europic

Tu Delft/EUROPIC Approach

Fundamentals of Process Intensification

Ind. Eng. Chem. Res. 2009, 48, 2465–2474 2465

100TH ANNIVERSARY REVIEW

Structure, Energy, Synergy, Time—The Fundamentals of Process Intensification

Tom Van Gerven and Andrzej Stankiewicz*

Process & Energy Department, Delft University of Technology, Leeghwaterstraat 44, 2628 CA Delft, The Netherlands

Spatial – Modifies the structure of the molecule, equipment or plant;
Thermodynamic – Applies more or different types of energy to effect an increase in fluid, mass, or heat transfer;
Synergy –Combining unit operations or making operations continuous; and
Time – Seeks to modify the structure of the equipment or a chemical to improve the limiting factors in the process

Example: Microreactor for manufacturing of a specialty product (DSM)

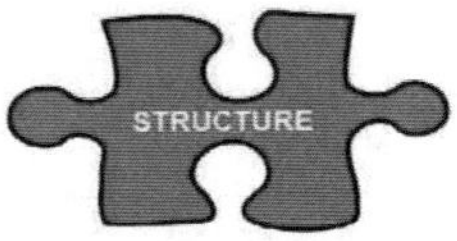

Traditional technology

Stirred Tank Reactor: the reactants are mixed in a large vessel, and the heat is removed through the jacket or a heat transfer coil.

PI technology

Microreactor: the reactants are mixed, and the heat is removed through thousands of micro channels, fabricated by micromachining or lithography

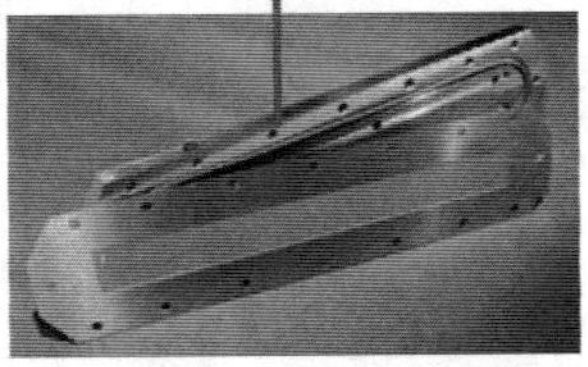

Benefits

- Equipment content 3 litres vs 10 m^3
- 20% higher selectivity → 20% higher material yield
- Process more reliable because continuous instead of batch
- Same capacity (1700 kg/h)

Photos courtesy of DSM and Forschungszentrum Karlsruhe

europic

DSM PI Methodology

The Function Oriented Approach

Put the requirements of the process at the top

- Start from the underlying elementary physical and chemical processes
- Analyze what functionalities (requirements are needed
- Identify resistances in individual process steps
- Design ideal pathway for molecules
- Then look at the design apparatus(or develop a new one) to achieve this

Dr. Jean-Piere Burnelle, Solvay Executive Vice President Process Innovation – PI is "Do more and better with less"

DSM has a core PI R&D group headed by a PI competency manager

The optimal pathway

Imagine a fluid element and a chemical conversion

$A + B \longrightarrow P \longrightarrow Z$

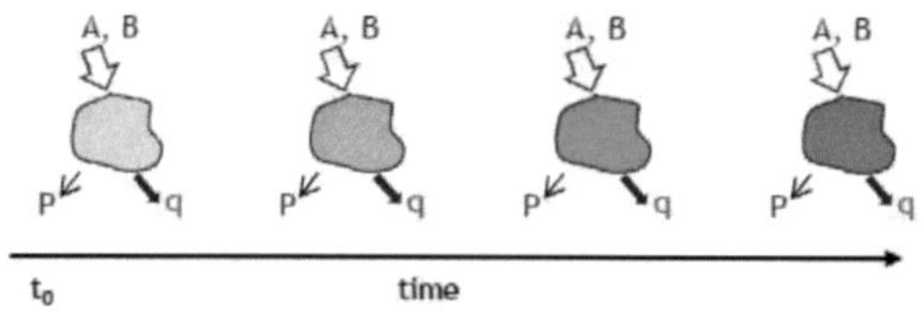

- Within the fluid element there are no gradients
- Chemical reaction → change of composition → change of optimal conditions
- We optimize conditions by manipulating the outer fluxes (heat, mass)
- If fluxes are not limited we will find the optimal pathway
- This optimum is independent of existing equipment

fit the equipment to the requirements

R. Reintjens

DSM

Example: Micro reactor in production

Objective: Manufacture the desired quality and yield at full GMP level.

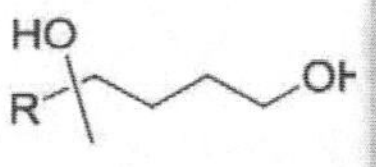

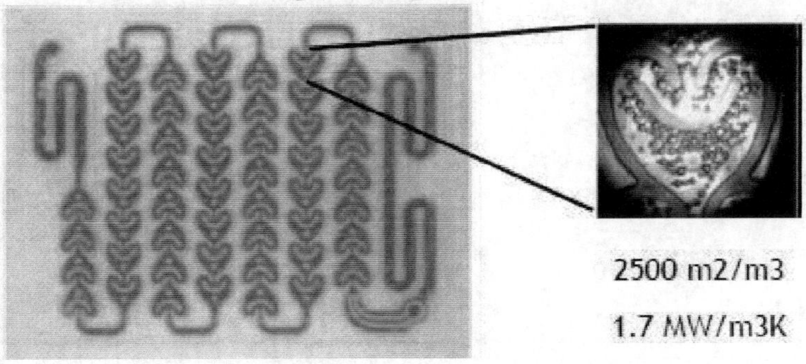

O_2

n

2500 m2/m3

1.7 MW/m3K

Process typ
- Batch ope
- Exotherm
- HNO_3 (hi
- Danger (r
- Low productivity (diluted)
- Unstable product

- **GMP production**

André H.M. de Vries

Challenge: Selectively get mono-nitro product

DSM BRIGHT SCIENCE. BRIGHTER LIVING.

Microreaction Technology in Production

- **Problems that Needed Solving**
  - Lab scale Development
  - Developing a Industrial Scale Microractor
    - *Pressure Drop*
    - *Mixing*
    - *Integration – Numbering up Problem*
  - Manufacturability of the reactor

DSM Full Scale Production

Key Lessons Learned

- **PI Hurdles**
  - should not be focused on reaction, but enlarged to downstream processing (separations…)
  - smaller doesn't mean cheaper…
  - easier to justify on new projects than on retrofit of existing units
  - A culture to be disseminated...

D. Horbez, Solvay

Findings

- PI has evolved from a "toolbox" of technologies approach to an **integrated multidisciplinary approach for understanding the relationships between fundamental science and process engineering**
- PI efforts start by asking the basic question: "**What is the limiting factor (rate, environmental, safety, capital cost, etc.) in the process or enterprise**?" This fundamental question was consistently brought up by all institutions visited;
- PI efforts involve **analyzing the underlying elementary physical and chemical processes with the goal of providing the optimal pathway for each molecule processed**;
- **All scales within an enterprise (molecular to plant to enterprise) should be considered** when applying PI;

Findings

- **Metrics for PI should be set to achieve a step change** in plant footprint, environmental release, capital/operating cost, or other metric of interest. PI disrupts cost paradigms and is not business as usual process optimization;
- A database and library of PI technologies are maintained by EUROPIC.
- **PI is a culture that needs to be disseminated to be effective**;
- The EU continues to fund large programs associated with PI and SM;
- Most Chemical Companies are involved in SMART Manufacturing and supply chain modeling efforts.

Spinning Disc Extractor – A Multistage Centrifugal Contactor in One Unit

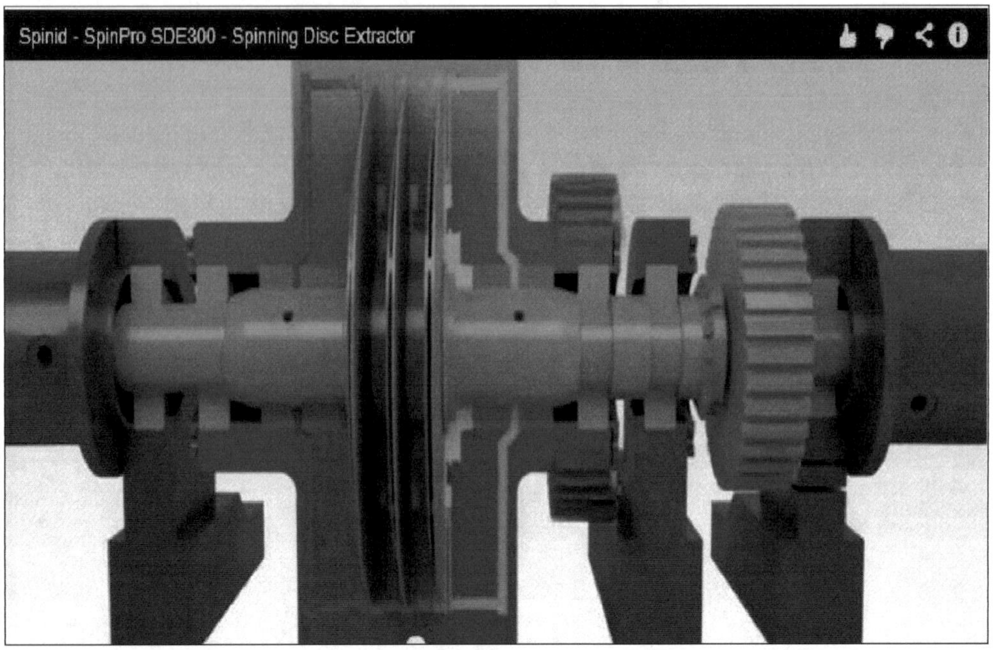

Closing Remarks

- **World Energy markets will change**
- **Innovation is the key to staying competitive**
- **A more efficient chemical industry is more sustainable**
- **And if that didn't motivate you …**

The Department of Energy is considering a Manufacturing Institute Associated with Process Intensification

Recently One Major Oil Company has devoted significant resources to Process Intensification

What is Your Company doing in this area?

Process Intensification for Energy Applications

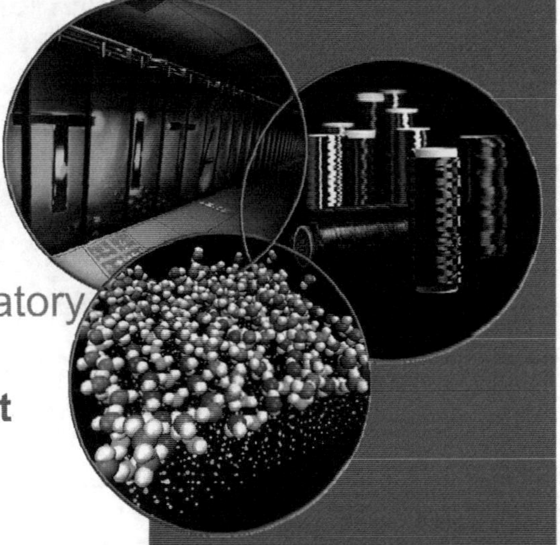

Costas Tsouris
Energy and Transportation Science Division
Oak Ridge National Laboratory

2015 Process Development Symposium
Emerging Trends in Process Research
June 2-4, 2015

ORNL is managed by UT-Battelle
for the US Department of Energy

OAK RIDGE National Laboratory

OUTLINE

- **Definitions**
- **Examples of PI Systems**
- **Examples from My Own Research**
- **Summary**
- **Role of Process Intensification in Advanced Manufacturing**

OAK RIDGE National Laboratory

Process Intensification (PI)/ High-Efficiency Modular Chemical Processing (HEMCP)

- **Engineering strategy of making dramatic reductions in the size of unit operations in order to reduce capital cost, footprint, inventory, energy requirements, and raw materials**
- PI/HEMCP Drivers
  - Novel or enhanced products
  - Improved chemistry
  - Enhanced safety
  - Improved processing
  - Energy and environmental benefits
  - Capital cost reduction
  - Diversified units – just-in-time philosophy (adjustable to market)
  - Enhanced corporate image

PI/HEMCP Reactors

- **High-G systems**
  - Spinning disc
  - Centrifugal contactor
- **Membrane catalytic reactors**
- **Heat-exchanger reactors**
- **Microreactors**
- **Oscillatory flow reactors**
- **Reactors/separators driven by external fields**
  - Electric
  - Magnetic
  - Sound
  - Gravity
  - Microwaves

Tsouris & Porcelli, *Chem. Eng. Prog.*, **99**(10), 50-55 (2003)
Process Intensification: Has Its Time Finally Come?

OAK RIDGE National Laboratory

Examples of PI/HEMCP Systems: Spinning Disc Reactor

- The multi-purpose, reactor developed by Colin Ramshaw, University of Newcastle
- It can be used for absorption, extraction, polymerization and other chemical processes

Fractal Technology

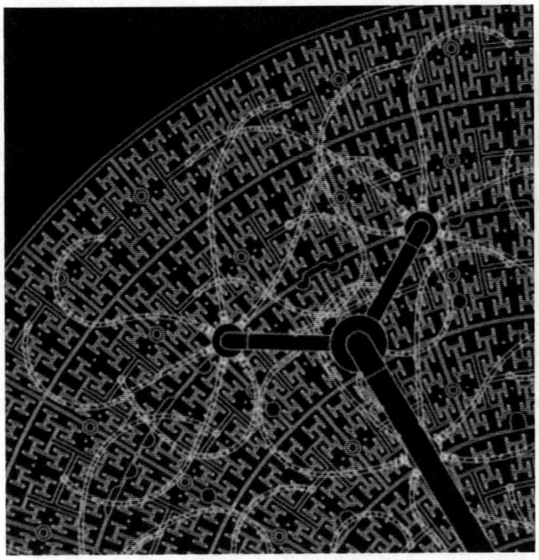

(http://www.arifractal.com/)

- Typical layout of a fractal used for fluid distribution to deliver or collect fluid
- The device resembles fluid transporting structures found in living organisms

Amalgamated Research Inc.

Fractal Technology in Ion Exchange

(http://www.arifractal.com/)

- Amalgamated Research Inc.'s fractal technology, used in ion exchange cell, increases resin productivity by a factor of 10
- Other fractal PI applications include chromatography, distillation, adsorption, aeration, extraction, and reactors

Amalgamated Research Inc.

Flex Reactor

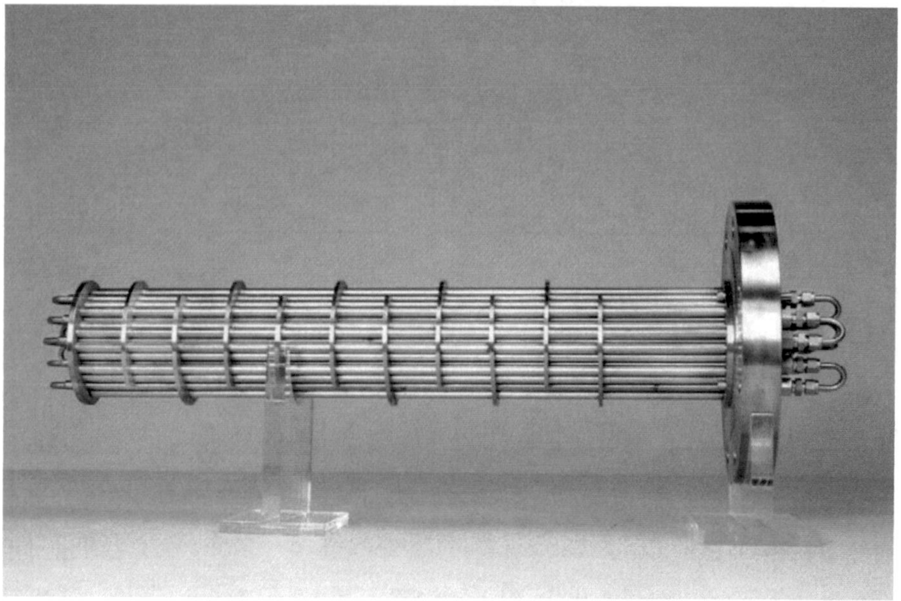

The FlexReactor (Flexible intensified reactor) developed by **BHR Group**
(http://www.pinetwork.org/pubs/cranfield/FlexReactorPINwebsite214-11-02.pdf)

On Site Process Intensification

On Site Process Intensification (OSPRI) plant: mobile PI plant designed by **BHR Group** to enable fine/specialty chemical companies to prove feasibility of PI processes at their site
(http://www.pinetwork.org/pubs/cranfield/FlexReactorPINwebsite214-11-02.pdf)

Microreactors

Example Compact MPT Reactor

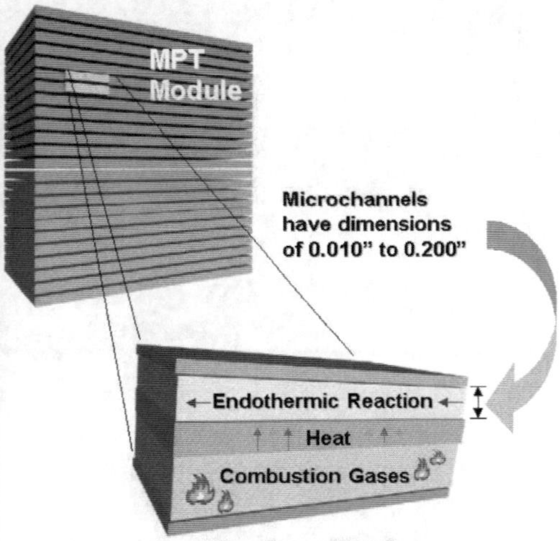

Velocys has developed manufacturing techniques to scale-up microreactors to industrial capacities

Velocys

Microreactors in Energy Applications

Fischer-Tropsch
FT microchannel reactors and superactive catalysts for the distributed production of liquid fuels through smaller scale gas-, biomass-, and coal-to-liquids applications

Innovative technology
Microchannel process technology and superactive catalysts that enable large scale economics for smaller scale plants

Velocys

Examples from My Group's Research:

EMR Reactor

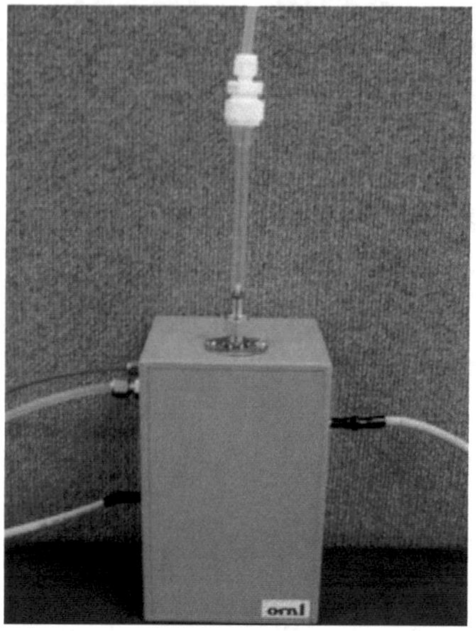

The Electrohydrodynamic Micromixing Reactor (EMR)
developed at the Oak Ridge National Laboratory (ORNL)

Electrohydrodynamic Micromixing Reactor

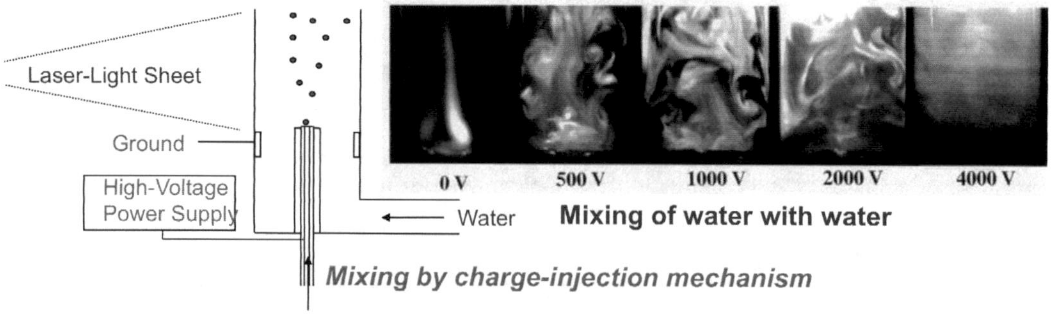

Mixing of water with water

Mixing by charge-injection mechanism

Mixing of butanol with water

Conditions: Butanol – 0.8 mL/min; Water – 50 mL/min; Tube ID = 7.5 mm

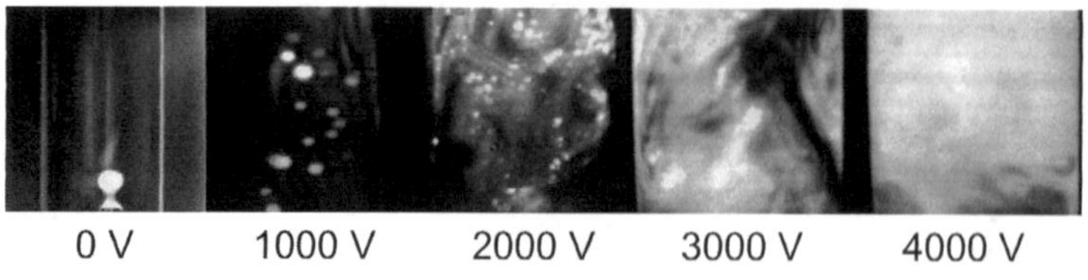

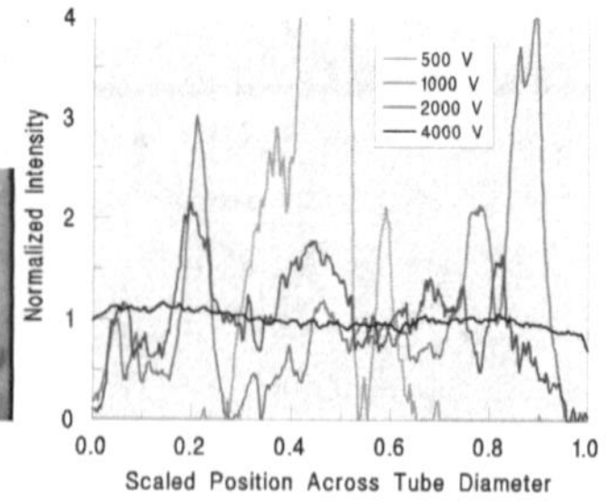

Applications of EMR in Particle Synthesis and Water Purification

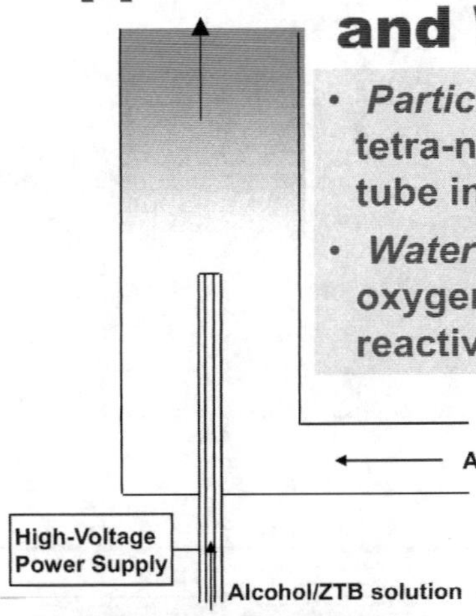

- *Particle synthesis*: Alcohol containing zirconium tetra-n-butoxide (ZTB) is injected through electrified tube into a flowing stream of alcohol-containing water
- *Water purification*: When ozone-containing gas or oxygen under discharge conditions is injected, highly reactive radical species are formed causing oxidation

Conditions for particle synthesis:
[ZTB] = 0.1 *M*
$[H_2O]$ = 0.5 *M* in butanol

Particles produced are more homogeneous at higher applied voltage

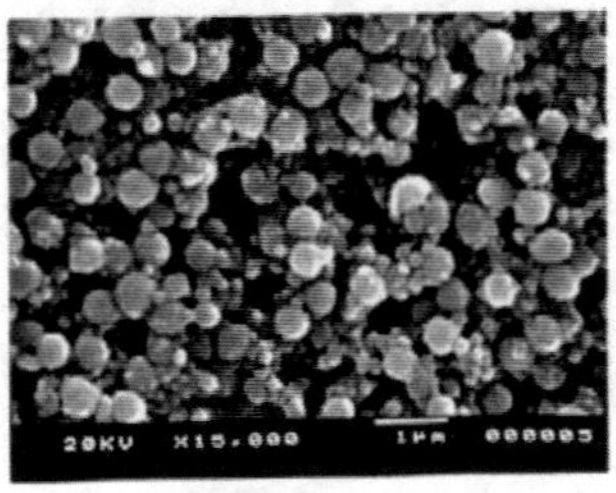

0 kV

5 kV

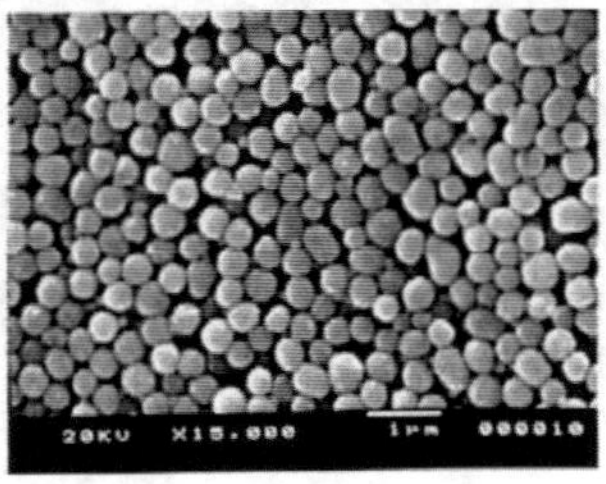

8 kV

Why EMR is a PI Process?

- Chemistry not working well by other means (e.g., stirring)
- Low-cost device
- Diversified reactor working with an electric field
  - Particle synthesis
  - Water purification
- Application demonstrated for microchannels and is possible for microreactors
- Concept better than that of a static mixer
  - Mixing in the delivery line
  - No moving parts
  - External control

ORNL Has a Long History in PI/HEMCP Devices Example: Centrifugal Contactor/Reactor

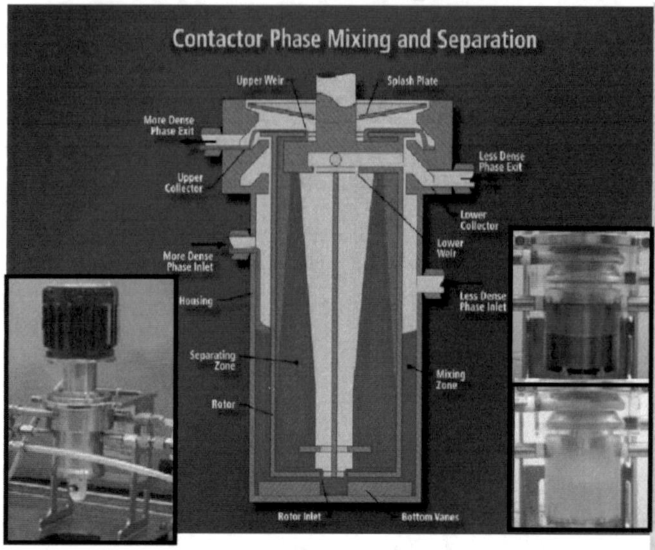

Developed for solvent extraction at DOE laboratories over 30 years ago and evolved from the mixer-settlers

Centrifugal Contactor

- Effective for mass transfer, reaction, and phase separation
- Demonstrated for liquid-liquid extraction
- In some cases, applicable for gas-liquid systems
- Reduces unit operations by integrating reaction/separation
- Provides intense mixing in the mixing zone and high G in the phase separation zone
- Allows high throughput or low residence time (seconds)
- Series of contactors can be used in a multistage operation
  - Continuous process
  - Effective process control
  - Scalable

Currently employed for high-level radioactive waste treatment at SRNL: A high-profile DOE application

OAK RIDGE National Laboratory

PI/HEMCP Applications Guided by Basic Studies

Experimental study for insight and calibration

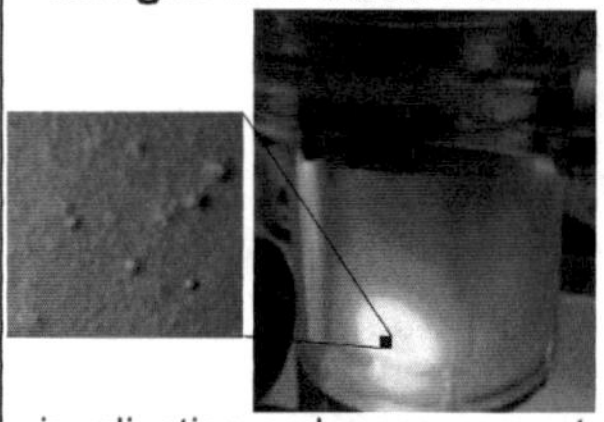

visualization and measurement

Sub-scale molecular modeling

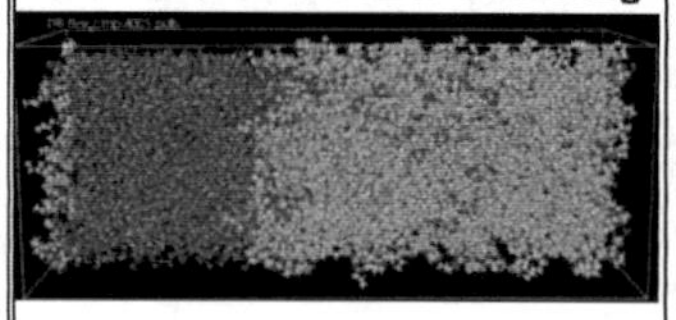

insight and transport parameters

Advanced Modeling and Simulation

- Multiphase 3D fluid flow
- Reactive multispecies

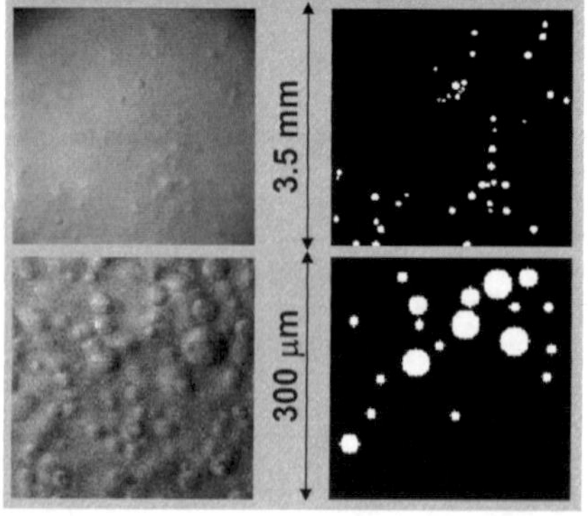

A coupled effort has been undertaken to provide realistic simulation of centrifugal contactors aimed at scale-up and novel designs

Modified Centrifugal Contactor to Control Residence Time

- A modified centrifugal contactor to allow control of the fluid residence time in the mixing zone
- The integrated reactor/separator can handle reactions requiring 1-2 minutes for completion
- Suitable for liquid-liquid reactive/extractive systems at temperatures up to 100°C or higher
- Allows addition of a reactive or inert gas
- Accumulated solids can be removed periodically
- Can be used for continuous biodiesel production

Modification was necessary to control residence time of the reactants

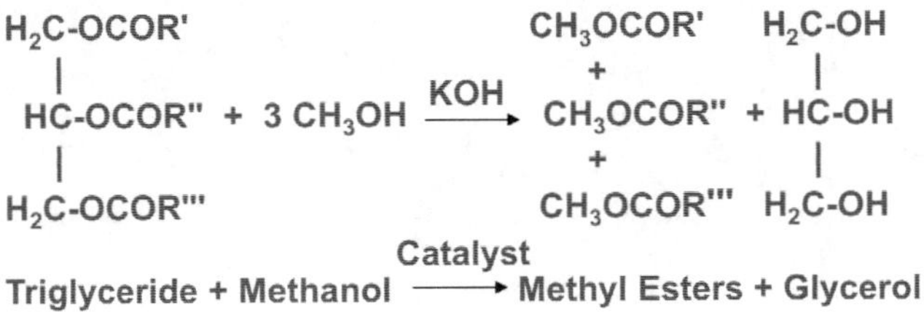

U.S. Patent # 8,097,219

Currently investigated for bio-oil separations

OAK RIDGE National Laboratory

Another Example of PI/HEMCP Reactor at ORNL: Continuous-Jet Hydrate Reactor (CJHR)

$$CO_{2(l)} + n\,H_2O_{(l)} \leftrightarrow CO_2 \cdot nH_2O_{(s)}$$

- Three-phase liquid/liquid/solid reactor or gas/liquid/solid reactor in the case of methane
- Formation of solid hydrate at the interface slows down the reaction significantly
- High conversion of gas into hydrate requires high surface area and effective removal of heat generated during hydrate formation reaction
- A continuous reactor maximizing the interfacial area between reactants (CO_2 and water or CH_4 and water) is needed

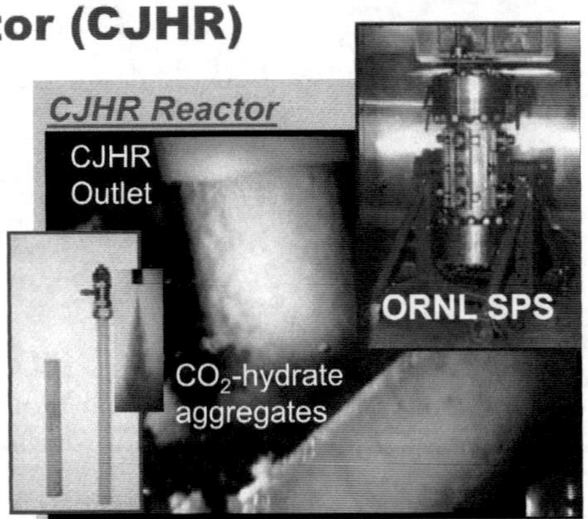

Successfully tested for CO_2 and CH_4 in the laboratory and in the ocean

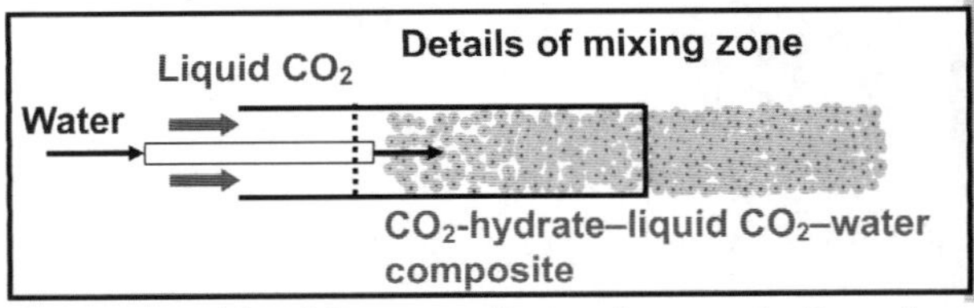

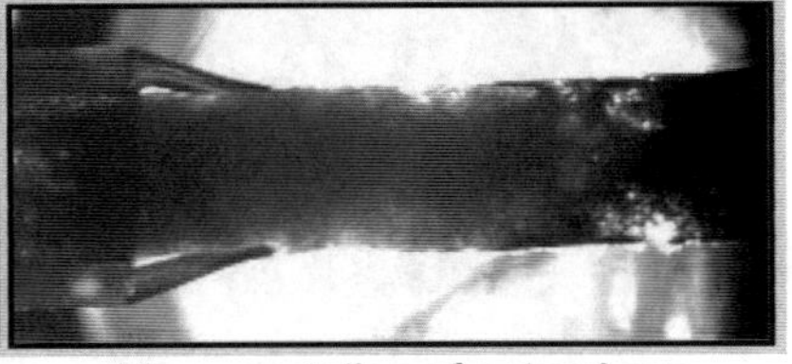

Can be used for CO_2 separation and sequestration, storage and transportation of natural gas, and produced water treatment

U.S. Patent # 6,598,407

OAK RIDGE National Laboratory

CO_2 Ocean Injection at Monterey Bay, CA

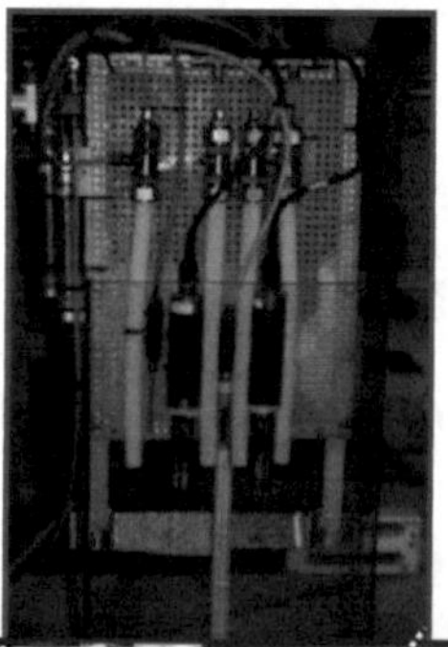

Tsouris et al, ES&T, 2004

Tsouris et al, Energy & Fuels, 2007

OAK RIDGE National Laboratory

Application of CJHR in Natural-Gas Storage and Transportation

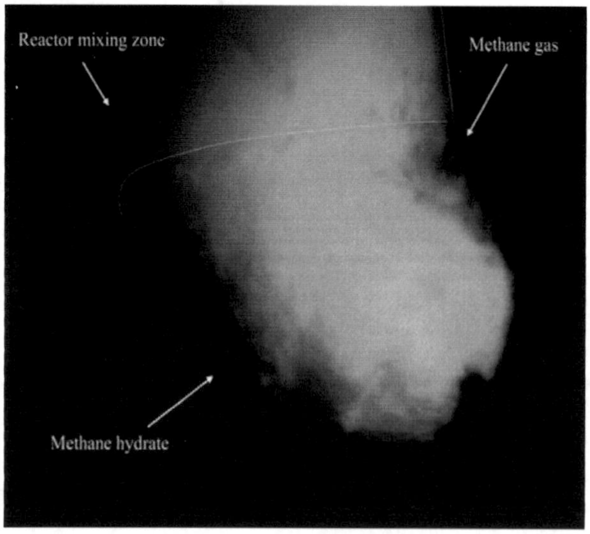

Taboada-Serrano et al, I&EC Res., 2009

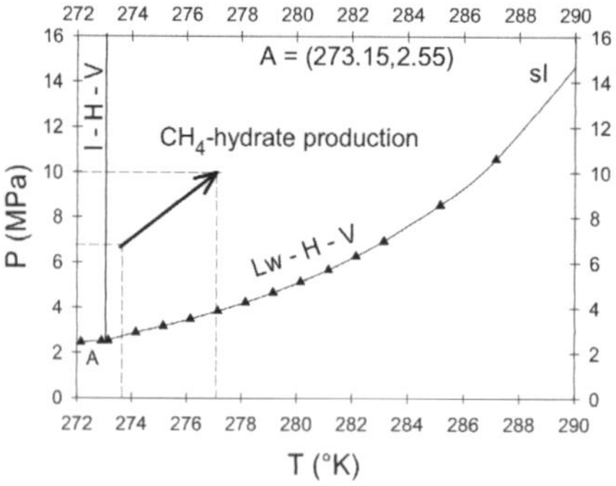

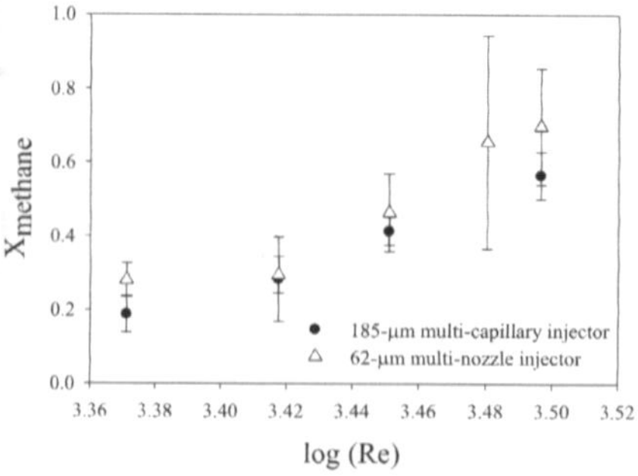

PI/HEMCP Development via Novel Materials

- PI/HEMCP devices can also be achieved via novel materials such as catalysts, solvents, membranes, sorbents, nanoparticles and nanostructures, fractal geometries, etc.
- Examples of novel materials at ORNL that help PI/HEMCP advances include:
  - Ionic liquids
  - Inorganic membranes
  - Nanostructured mesoporous carbon

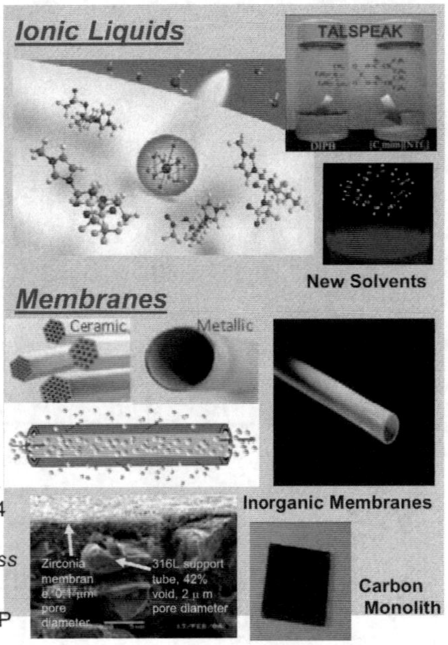

ORNL ionic liquids employed for tough separations problems:

Performance of Several Ionic Liquids for the Separation of 1-Octene from *n*-Octane, *Ind. Eng. Chem. Res.* 2009, *48*, 11168–11174
Gary J. Belluomini,* John G. Pendergast, Christopher H. Domke, and Bryson R. Ussing *The Dow Chemical Company, Engineering & Process Sciences Laboratory, Texas Operations, Freeport, Texas 77541*

"...ORNL ionic liquid showed selectivities 24-32 times higher than NMP while maintaining a comparable mass capacity as the nonionic solvent... has the potential to be used in an economically viable commercial process to recover valuable olefinic components from streams where they are mixed with paraffins."

OAK RIDGE National Laboratory

Summary on PI Development

- PI offers several advantages for industry and society
- Follows principles of sustainable development
- Leads to cost reduction by "doing more with less"
- Reduces size and number of unit operations by combining units
- Allows new chemistry, not possible by existing reactors
- Safer and environmentally advanced approach
- Small, versatile operations are possible, reducing the transportation cost of feedstocks
- Leads to faster operations with less footprint and lower chemical inventories
- There are also barriers to PI
  - Difficult to convince mature industry
  - Limited current portfolio of PI processes
  - New design codes needed for novel equipment
  - Inadequate education
- R&D should proceed together with education

OAK RIDGE National Laboratory

Why PI/HEMCP Development Is Currently an Important Area?

- PI/HEMCP concept can play a strong role for untapped feedstocks such as natural gas, biomass, and mine tailings
- It can play a role in energy systems, specialty chemicals & polymers, waste & recycling, mining & refining
- PI/HEMCP will lead to advanced manufacturing opportunities
  - 3-D printing, additive manufacturing
- Current advances in manufacturing can also boost the development of smart PI/HEMCP devices
  - 3-D printing of mixer/settler systems
- The link between PI/HEMCP development and advanced manufacturing has been recognized

PI/HEMCP drivers are expected to push for further advances that will boost manufacturing. Advance manufacturing will further enhance PI/HEMCP with smart devices

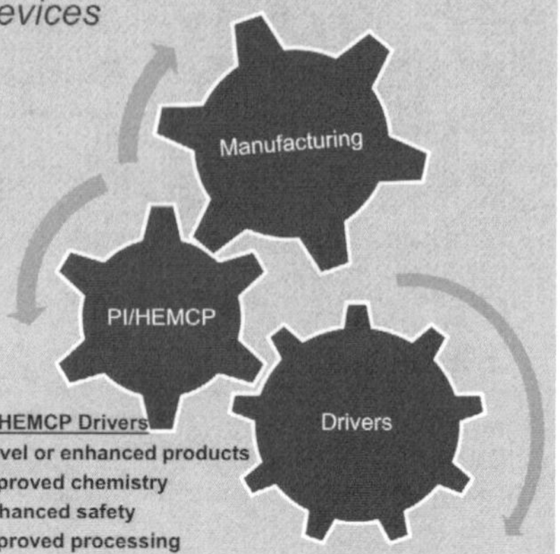

PI/HEMCP Drivers
Novel or enhanced products
Improved chemistry
Enhanced safety
Improved processing
Energy and environmental benefits
Capital cost reduction
Enhanced corporate image
Diversified units

OAK RIDGE National Laboratory

Acknowledgments

This work was sponsored by the US Department of Energy, under contract DE-AC05-00OR22725 with Oak Ridge National Laboratory, managed by UT-Battelle, LLC.

Thank You!

OAK RIDGE National Laboratory

Case Study: Debottlenecking a Solid Liquid Isolation: Laboratory to Capital Project

Tomazi, K., Mallinckrodt

This work is a case study in the development efforts required to effectively de-bottleneck a bulk API intermediate manufacturing process. The goal was to increase process capacity within the constraints of quality, timeline, manufacturing footprint space and capital constraints. The existing solid-liquid isolation step was originally done using a batch centrifuge that was shared with a subsequent step in the process. Development work was done to optimize the upstream crystallization step to improve performance of current operations and create the necessary critical physical property data and filtration mechanisms to drive the selection process to identify the best technology. This work identified the indexing belt filter as the best technology. A pilot study was done with an indexing belt filter to verify proof of concept. This work revealed significant cake cracking occurred, resulting in non-uniform deliquoring and cake washing that led to the degradation quality of the isolate solids. Further development work was required to identify the necessary conditions required to avoid cake cracking. A second round of pilot studies was done on a larger scale that provided both proof of the generation of quality wet cake and the necessary design data for large scale purchase. The end result is the generation of a new isolation process, a semi- continuous filtration that increased capacity 84%, created less handling and better upstream and downstream process flow.

"INTRAPRENEURSHIP": HOW TO MANAGE NEW PROCESS DEVELOPMENT TO MAXIMIZE VALUE

AIChE Process Development Symposium,
2-4 June 2015, Houston, TX

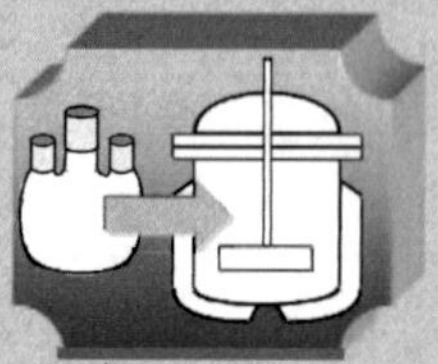

AIChE Process Development Division-12 www.pd-aiche.com

Joe Powell [Joseph B. Powell, Ph.D.]
Shell Chief Scientist - Chemical Engineering

SIEP 1

DEFINITIONS AND CAUTIONARY NOTE

The companies in which Royal Dutch Shell plc directly and indirectly owns investments are separate entities. In this presentation "Shell", "Shell group" and "Royal Dutch Shell" are sometimes used for convenience where references are made to Royal Dutch Shell plc and its subsidiaries in general. Likewise, the words "we", "us" and "our" are also used to refer to subsidiaries in general or to those who work for them. These expressions are also used where no useful purpose is served by identifying the particular company or companies. ''Subsidiaries'', "Shell subsidiaries" and "Shell companies" as used in this presentation refer to companies over which Royal Dutch Shell plc either directly or indirectly has control. Companies over which Shell has joint control are generally referred to "joint ventures" and companies over which Shell has significant influence but neither control nor joint control are referred to as "associates". In this presentation, joint ventures and associates may also be referred to as "equity-accounted investments". The term "Shell interest" is used for convenience to indicate the direct and/or indirect (for example, through our 23% shareholding in Woodside Petroleum Ltd.) ownership interest held by Shell in a venture, partnership or company, after exclusion of all third-party interest.

This presentation contains forward-looking statements concerning the financial condition, results of operations and businesses of Royal Dutch Shell. All statements other than statements of historical fact are, or may be deemed to be, forward-looking statements. Forward-looking statements are statements of future expectations that are based on management's current expectations and assumptions and involve known and unknown risks and uncertainties that could cause actual results, performance or events to differ materially from those expressed or implied in these statements. Forward-looking statements include, among other things, statements concerning the potential exposure of Royal Dutch Shell to market risks and statements expressing management's expectations, beliefs, estimates, forecasts, projections and assumptions. These forward-looking statements are identified by their use of terms and phrases such as ''anticipate'', ''believe'', ''could'', ''estimate'', ''expect'', ''goals'', ''intend'', ''may'', ''objectives'', ''outlook'', ''plan'', ''probably'', ''project'', ''risks'', "schedule", ''seek'', ''should'', ''target'', ''will'' and similar terms and phrases. There are a number of factors that could affect the future operations of Royal Dutch Shell and could cause those results to differ materially from those expressed in the forward-looking statements included in this presentation, including (without limitation): (a) price fluctuations in crude oil and natural gas; (b) changes in demand for Shell's products; (c) currency fluctuations; (d) drilling and production results; (e) reserves estimates; (f) loss of market share and industry competition; (g) environmental and physical risks; (h) risks associated with the identification of suitable potential acquisition properties and targets, and successful negotiation and completion of such transactions; (i) the risk of doing business in developing countries and countries subject to international sanctions; (j) legislative, fiscal and regulatory developments including regulatory measures addressing climate change; (k) economic and financial market conditions in various countries and regions; (l) political risks, including the risks of expropriation and renegotiation of the terms of contracts with governmental entities, delays or advancements in the approval of projects and delays in the reimbursement for shared costs; and (m) changes in trading conditions. All forward-looking statements contained in this presentation are expressly qualified in their entirety by the cautionary statements contained or referred to in this section. Readers should not place undue reliance on forward-looking statements. Additional risk factors that may affect future results are contained in Royal Dutch Shell's 20-F for the year ended December 31, 2012 (available at www.shell.com/investor and www.sec.gov). These risk factors also expressly qualify all forward looking statements contained in this presentation and should be considered by the reader. Each forward-looking statement speaks only as of the date of this presentation, 01 June 2015. Neither Royal Dutch Shell plc nor any of its subsidiaries undertake any obligation to publicly update or revise any forward-looking statement as a result of new information, future events or other information. In light of these risks, results could differ materially from those stated, implied or inferred from the forward-looking statements contained in this presentation.

We may have used certain terms, such as resources, in this presentation that United States Securities and Exchange Commission (SEC) strictly prohibits us from including in our filings with the SEC. U.S. Investors are urged to consider closely the disclosure in our Form 20-F, File No 1-32575, available on the SEC website www.sec.gov. You can also obtain these forms from the SEC by calling 1-800-SEC-0330.

SIEP 2

PROCESS DEVELOPMENT: CORE COMPETENCY FOR INDUSTRY

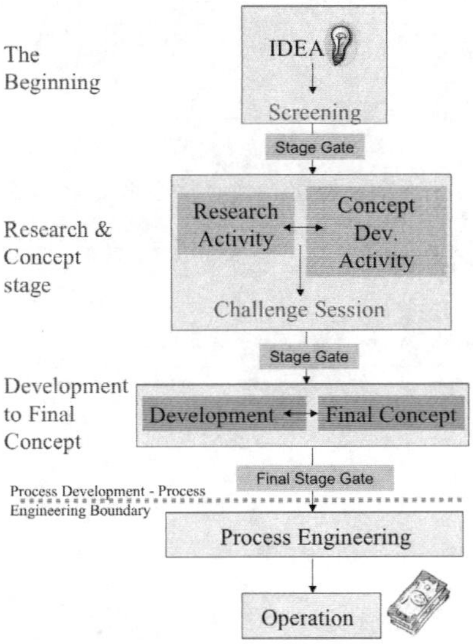

Ability to screen for viability before spending huge $$$

Ability to deliver processes that work

Ability to anticipate business needs

SIEP

3

ELEMENTS OF PROCESS DEVELOPMENT

Stage-gate approach, with iteration

- Preliminary screening for technical and economic viability
  - Establishment of business case and objectives
- Process studies
- Scale-up tests for key unit operations
- Fully integrated demonstration
- Commercial design package
- Startup support

Multi-disciplined Team

- Process chemist
- Process development engineer
- Process economics / evaluations engineer

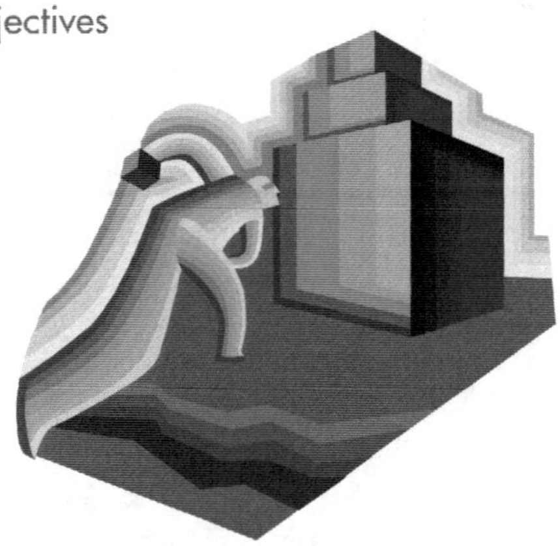

SIEP

4

TECHNOLOGY INNOVATION AND DEVELOPMENT

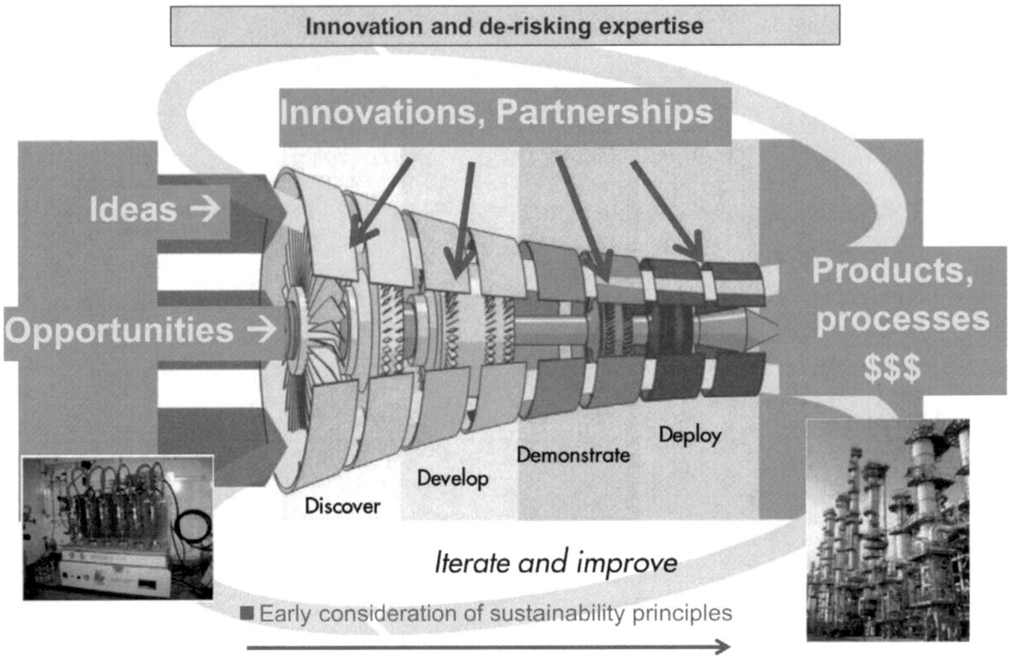

Process Development: Differentiator for Industry

- Academia or startup companies usually lack process development skill, experience and resources.
- Is a key component of what industry brings to the table in partnerships for technology development.
- Will make or break a program:
  - New discovery (IP) and critical know how are invariably generated *during* process development.

PROCESS ENGINEERING AND EVALUATIONS

$$\frac{C_J}{C_i} = \left[\frac{V_J}{V_i}\right]^{0.6}$$

PLANT SIZE IMPACT

Unit capex costs decrease as plant size increases basis market growth / risk.

Everpresent "learning" factor from basic chemical engineering plant design power law.

Simulation and Optimization

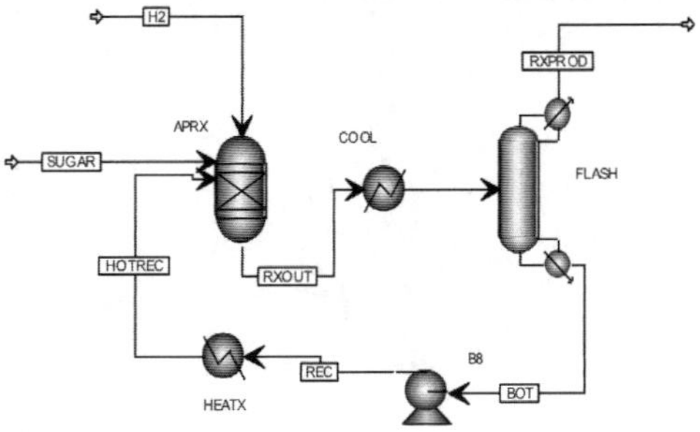

SIEP 7

Development Cost vs. R&D Stage

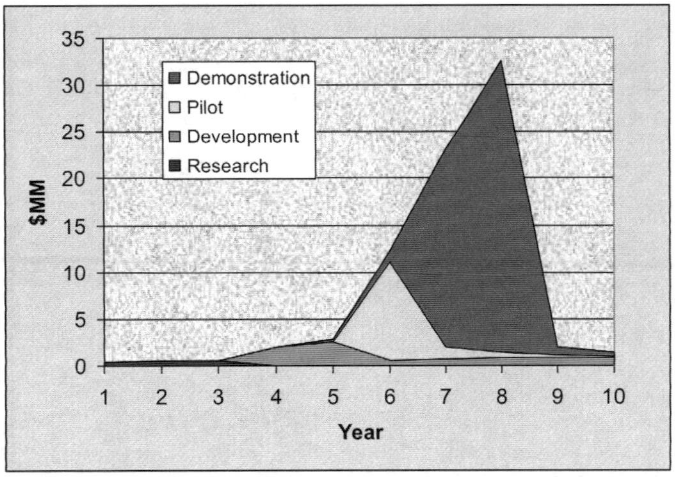

- "Research is cheap; Development is expensive!"
- High cost of demonstration units.

TECHNOLOGY READINESS LEVELS

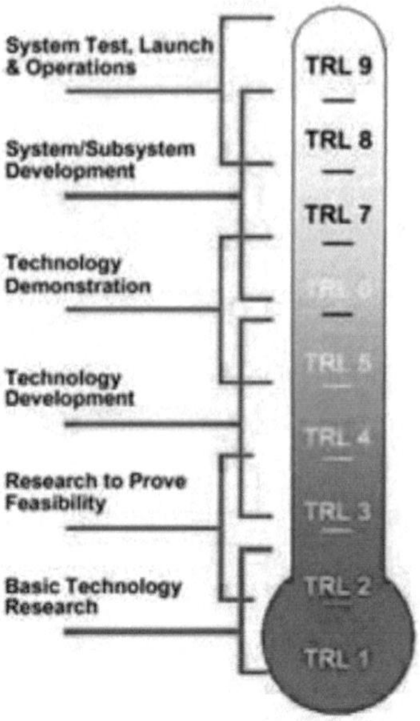

Metrics for assessing maturity.

NASA (from www.wikipedia.org)

SIEP 9

SUSTAINABLE DEVELOPMENT: SHELL AS A LEADER AND DRIVER FOR INNOVATION

"Meeting the needs of the present generation without compromising the ability of future generations to meet their needs."

People, Planet, and Profits

Sustainability is more than just CO_2

Water
Land Use
Biodiversity
Managing Chemicals
Social performance
Security / jobs
Other environmental stressors
Safety
Climate and CO_2

Key challenges: people, systems, priorities

- People and Systems can be as challenging as technology
  - Communications key to managing.
- Timing:
  - Projects are ultra urgent or not needed.
  - Business favors short term focus
- Need advocate for strategic technology; R&D
  - Keep portfolio of viable projects
- Getting things done
  - Systems compliance & productivity

Managing Innovation: Risk Appetite

Incremental (evolutionary) vs. revolutionary must be managed relative to current economics.

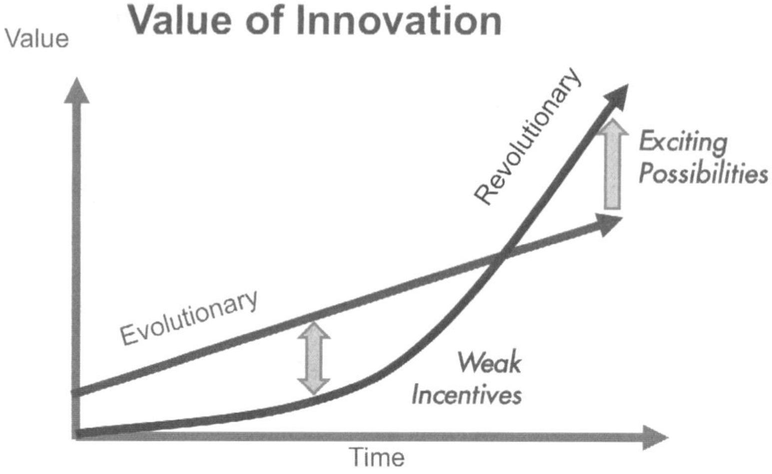

DERISKING TECHNOLOGY BY EXPLICIT ASSUMPTION

- Alternative to "we'll just learn our way to better economics"
  - Aka learning or experience curves

20% cost reduction per doubling in capacity <u>for known process technology</u>

$$C = k \bullet Q^{-\alpha}$$
$$k > 0$$
$$\alpha > 0$$

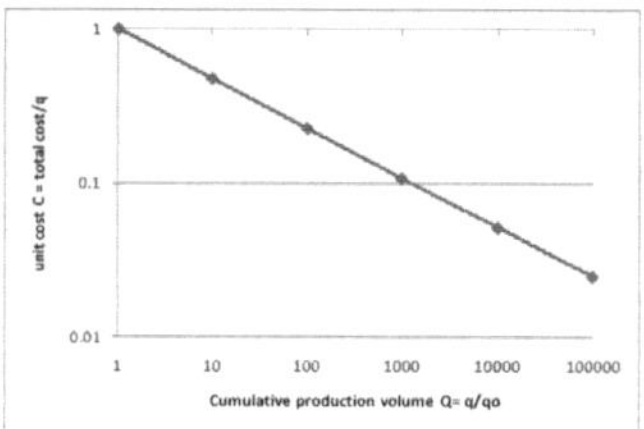

- Explicit process / project economics improvement potential (best possible):
  1. Catalyst life
  2. Reaction kinetics
  3. Separation efficiency
  4. Process integration / simplification
  5. Heat integration

- Scale-up / market size (power law economies)
- Alternate feedstocks
- Changes in feed or energy costs

14

Psychology 101: helping business & management cope

Perspective:

Don't stop program until stop learning. Some "facts" or paradigms are generally wrong.

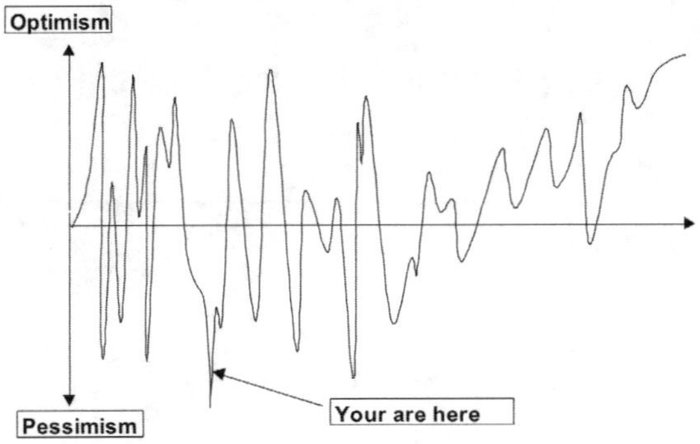

PROCESS DEVELOPMENT CAREER OPPORTUNITIES

- Growth (Technical, Business)
- Networking / Organizational knowledge
- Leverage
- Insight on "pulse" of organization for which technology is a differentiating factor.

No need to change career path

- The "opportunities" find you!

Growth by "filling in the white spaces"

- Defined as needs where no "expert" is available.
- Business customers will pay for your training
- Knowledge gained can be leveraged into future applications
- Opportunity to extend skill set & capabilities

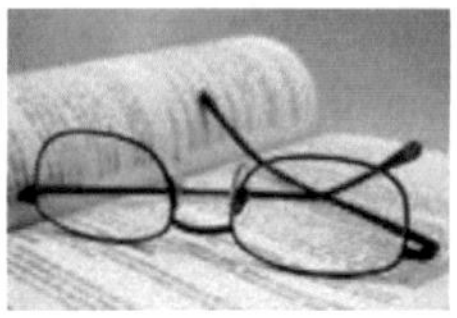

Innovation in Shell: multiple routes

- http://sww.shell.com/innovation/in_shell/hunters_network.html
- https://eu001-sp.shell.com/sites/AAAAA7456/default.aspx

To become the most **competitive** and **innovative** energy company in the world...

Hunters Network

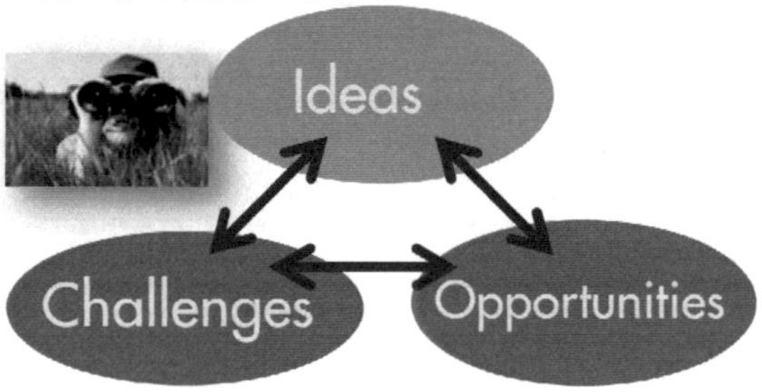

Shell

GameChanger

How it works...
When you have a GameChanging idea

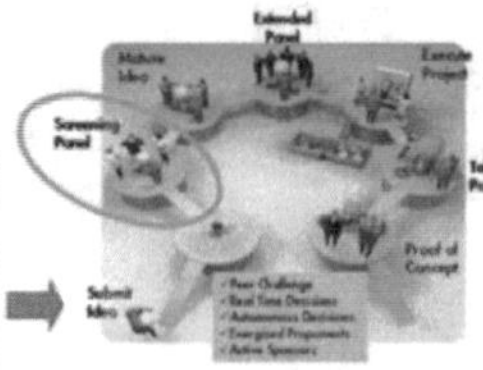

- Business Units & sponsors
- Emerging Technologies

Process Development Examples

- Enhanced Oil Recovery (energy crisis)
- Commercial Process "Fix" (after start-up failure)
- New Oxidation Process ("must-have" business)
- New Process and Product co-development
  - ("Must-have" new core business to justify R&D center)
- Alternative Energy Biofuels (new foundation)

Personal and Process Safety: fast track and ALARP

- Process and personal safety is responsibility of process development lead.
  - Norms are developed during piloting
  - Safe operation is fastest track!
- ALARP: Min- and Micro-scale experiments Chemistry happens at nano or angstrom scale!
  - Test quickly in existing apparatus!
  - Reduced risk & faster deployment
- Simulation and modeling can be used for scaleup.
- Use of mock fluids with lower hazards for faster testing of fluid flow, mixing, and hydrodynamic issues.

SEVERITY	CONSEQUENCES				INCREASING LIKELIHOOD				
	People	Assets	Environment	Reputation	A Never heard of in the Industry	B Heard of in the Industry	C Has happened in the Organisation or more than once per year in the Industry	D Has happened at the Location or more than once per year in the Organisation	E Has happened more than once per year at the Location
0	No injury or health effect	No damage	No effect	No impact					
1	Slight injury or health effect	Slight damage	Slight effect	Slight impact					
2	Minor injury or health effect	Minor damage	Minor effect	Minor impact					
3	Major injury or health effect	Moderate damage	Moderate effect	Moderate impact					
4	PTD or up to 3 fatalities	Major damage	Major effect	Major impact					
5	More than 3 fatalities	Massive damage	Massive effect	Massive impact					

WHAT DOES A BOWTIE LOOK LIKE?

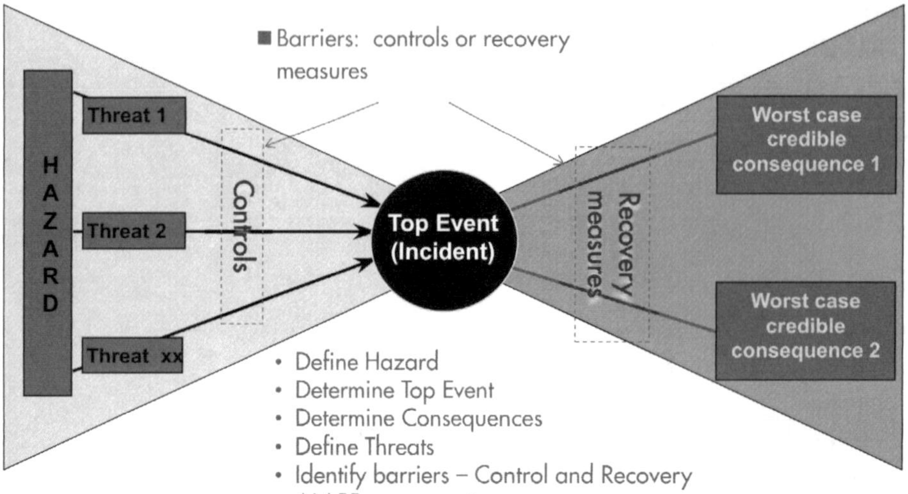

- Define Hazard
- Determine Top Event
- Determine Consequences
- Define Threats
- Identify barriers – Control and Recovery
- ALARP assessment

→*Take ownership of Process Safety, from Lab to Commercial!*

Process Development: Key Messages for Intrapreneur

- Business / economics mind set and drivers
- Champion vs. realistic appraisal (balance)
  - Resource and risk management: would you invest?
- Technology development:
  - Own all aspects through commercialization & deployment
  - Fit-for-purpose solutions
  - Superimposed on key technology differentiators
  - Lifelong learning & toolkit development
- Process safety owner and personal safety leadership

THANK YOU!

JOE.POWELL@SHELL.COM

DOWNRIGHT PROJECT MANAGEMENT

Nareshkumar Handagama,
PhD, PE, FAIChE, FIChemE (UK), C.Eng.

June, 2015

CHEMISTRY THAT MATTERS™

ESSENTIALS OF PROJECT MANAGEMENT

- The management of an engineering project is too important to be left to chance
- Project management is about creating an environment and conditions, or a culture in which a defined goal or objective can be achieved in a controlled (timely/ & efficient) manner by a team of people
- If the Journey is lost you'll never reach the Destination

Most often the managers are too intensely focused on the Destination and as a result lose sight on the Journey

No. 1

DOWNRIGHT PROJECT MANAGEMENT

(Definition of) Project Management ...

- ...is an important task
- ...has a definite beginning and end
- ...track tasks and measure accomplishments
- ...utilize quantitative "yard-Sticks", Work Breakdown Structures, Gantt and PERT charts
- ...need resources on an *ad-hoc* basis (as opposed to organizations)
- ...goal is to reduces risk and increases the chance of success

The management of an engineering project is too important to be left to chance.

No. 2

THE TRIPLE CONSTRAINTS

- Projects must be within cost
- Projects must be delivered on time
- Projects must be within scope

Having a management plan will provide the control to shape the project as desired: Cost, Scope, and Time

No. 3

EXPECTATIONS (PROJECT MANAGEMENT) DIAMOND

- Project within projected Cost
- Projects Delivery on time
- Projects within Scope of practice
- Project is of high Quality

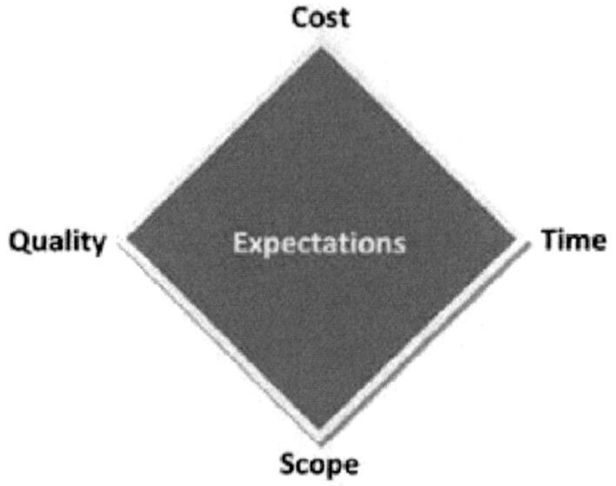

Engineering Project Planning with Quality

No. 4

THE SIX KEY PHASES OF A PROJECT

Project Phases

- **Definition:** Defining the goals, objectives and critical success factors
- **Initiation:** Everything that is needed to set-up the project before work can start
- **Planning:** Detailed plans of work including time, cost and resource estimates
- **Execution:** Work to deliver the product, service or desired outcome.
- **Monitoring & Control:** Ensure project stays on track with corrective action
- **Closure:** Formal acceptance of deliverables and disbanding all activity

Develop a management plan for a particular project consistent with the manager's mission and philosophy

No. 5

THE ROLE OF THE PROJECT MANAGER

Important Activities of the Project Manager

- Define Project - manageable sub-tasks, obtain resources and build a team
- Set Final Project Goals - motivate the team to complete the project on time
- Inform all stakeholders of progress on a regular basis
- Assess and monitor risks to the project and mitigate them
- No project ever goes exactly as planned – adapt, correct and manage

Develop an individualized management plan that works for a particular project

No. 6

PROJECT MANAGER'S MUST HAVE SKILLS

A Project Manager's Essential Skills

- Leadership – what is your leadership style?
- People management and people skills (balance competition vs. cooperation)
- Effective communication - verbal and written – including active listening
- Influencing and Negotiation
- Conflict management and resolution
- Planning
- Contract management and Time/Resource Estimation
- Problem solving and Creative thinking
- Time management
- Decision making skills

Formulate a management model that will best fit the project to accomplish the goals

No. 7

THINGS CAN SURELY GO WRONG IN A PROJECT

Possible Barriers and Pitfalls

- Poor communication
- Disagreements
- Misunderstandings
- Bad weather
- Union strikes
- Personality conflicts
- Poor management
- Poorly defined goals and objectives
- An environment not conducive to productivity

Define the relationships among administrative staff, engineering staff and the technical crew for effective operations the project management construct

No. 8

BENEFITS OF GOOD PROJECT MANAGEMENT

A good project management discipline will not eliminate all risks, issues and surprises, but will **anticipate** issues and provide processes/procedures to deal with important issues

- Projects finishing late, exceeding budget or not meeting customer expectations
- Inconsistency between the processes and procedures used by projects managers
- Successful projects, despite a lack of planning, achieved through high stress levels, goodwill and significant amounts of overtime
- Project management seen as not adding value and as a waste of time and money.
- Unforeseen internal and/or external events impacting the project.

Construct comprehensive policies/procedures for the five essential areas: people; money; supplies & equipment; activities; and communication channels

No. 9

PROJECT PLANNING

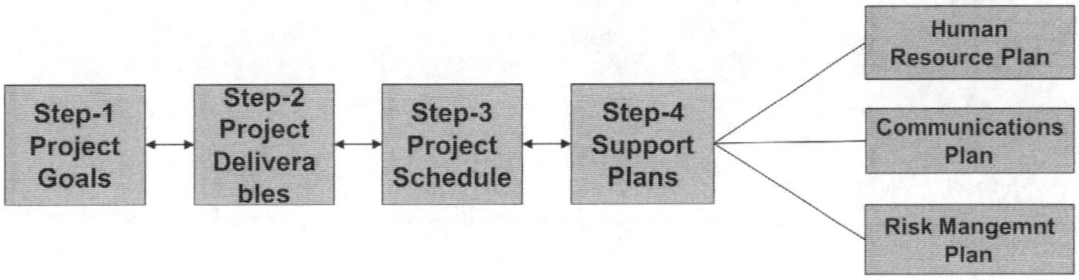

The key to a successful project is in the planning. Creating a project plan is the first thing you should do when undertaking any kind of project

No. 10

STEP 1: PROJECT GOALS

- A project is successful when the needs of the stakeholders have been met
- A stakeholder is anybody directly, or indirectly impacted by the project
- It is important to identify the stakeholders in your project

Examples of Stakeholders

- The project sponsor
- The customer who receives the deliverables
- The users of the project outputs
- The project manager and project team

Establish goals for the project plan with the needs and expectations of stakeholders

No. 11

STEP 2: PROJECT DELIVERABLES

Once the goals are defined in step 1

- Formulate a list of items as deliverables in order to meet those goals
- Specify what, when and how each item must be delivered
- State the deliverables in the project plan
- Specify an estimated delivery date for each deliverable

Accurate delivery dates will be established during the scheduling phase

Defined deliverables will qualify each goal

No. 12

STEP 3: PROJECT SCHEDULE

For each task identify

- Amount of effort (hours or days) required to complete the task
- The resource/s who will carryout the task
- Update deliverables with accurate delivery dates

Update with feedback

- Renegotiate the deadline (project delay)
- Employ additional resources (increased cost)
- Reduce the scope of the project (less delivered)
- Use the project schedule to justify pursuing one of these options

Create a list of tasks that need to be carried out for each deliverable identified in step 2
Reevaluate with feedback

No. 13

STEP 4: SUPPORTING PLANS

➢ **Human Resource Plan**
- Identify individuals and organizations with a leading role in the project
- Describe their roles and responsibilities on the project

➢ **Communications Plan**
- Create a document showing who needs to be kept informed about the project and how they will receive the information
- The most common mechanism is a weekly or monthly progress report, describing how the project is performing, milestones achieved and work planned for the next period

➢ **Risk Management Plan**
- Risk management is an important part of project management. It is important to identify as many risks to your project as possible, and be prepared with mitigation and contingency plan.

No. 14

EXAMPLES OF COMMON PROJECT RISKS

- Too optimistic time and cost estimates
- Poorly scheduled customer review and feedback cycle
- Unexpected budget cuts
- Unclear roles and responsibilities
- Lack of active listening resulting in lack of stakeholder or their needs not properly met
- Stakeholders changing requirements after the project has started
- Stakeholders adding new requirements after the project has started
- Poor communication resulting in misunderstandings, quality problems and rework
- Lack of resource commitment

Remember! Ignoring risks is not an option.
They don't go away

No. 15

sabic

SUMMARY

The Project Plan

- **Why?** - The problem and what is to be address by the project
- **What?** - The work to be done and the deliverables
- **Who?** - Personnel and what will be the responsibilities how are they organized
- **When?** - The project timeline and the milestones to be completed
- **Where?** – All this is to take place

The objective of a project plan is to define the approach to be used by the Project team to deliver the intended project management scope of a particular project.

No. 16

sabic

A GOOD PROJECT MANAGER IS A "SERVANT LEADER"

The Seven Pillars of Servant Leadership

- Servant-leaders are individuals of character
(i.e., Values diverse opinions; Cultivates a culture of trust; Develops other leaders; Helps people with life issues; Encourages; Sells instead of tells; Other focused; Thinks long-term; Acts with humility)
- Servant- leaders are skilled communicators
- Servant-leaders are compassionate collaborators
- Servant-leaders use foresight
- Servant-leaders are systems thinkers
- Servant-leaders exercise moral authority

- Communication
- Empathy
- Experience

A true leader will never demand respect but will always command respect

No. 17

ساىك
sabic

CONCLUSION

- When you have a management plan that seems right for your specific project, you've completed a necessary step on the road to effective action
- There are NO "short-cuts" to experience
- Keep your calm when all about you are losing theirs and blaming it on you
- Don't worry about your fall, instead find out the reason why you slipped

In a project the **Destination** is of cardinal importance, however every step of the **Journey** is paramount

"We must be silent before we can listen. We must listen before we can learn. We must learn before we can prepare. We must prepare before we can serve. We must serve before we can lead." ~*William Arthur Ward*

No. 18

ساىك
sabic

DISCLAIMER

DISCLAIMER: THE MATERIALS, PRODUCTS AND SERVICES OF SAUDI BASIC INDUSTRIES CORPORATION (SABIC) OR ITS SUBSIDIARIES OR AFFILIATES ("SELLER") ARE SOLD SUBJECT TO SELLER'S STANDARD CONDITIONS OF SALE, WHICH ARE AVAILABLE UPON REQUEST. INFORMATION AND RECOMMENDATIONS CONTAINED IN THIS DOCUMENT ARE GIVEN IN GOOD FAITH. HOWEVER, SELLER MAKES NO EXPRESS OR IMPLIED REPRESENTATION, WARRANTY OR GUARANTEE (i) THAT ANY RESULTS DESCRIBED IN THIS DOCUMENT WILL BE OBTAINED UNDER END-USE CONDITIONS, OR (ii) AS TO THE EFFECTIVENESS OR SAFETY OF ANY DESIGN OR APPLICATION INCORPORATING SELLER'S MATERIALS, PRODUCTS, SERVICES OR RECOMMENDATIONS. UNLESS OTHERWISE PROVIDED IN SELLER'S STANDARD CONDITIONS OF SALE, SELLER SHALL NOT BE RESPONSIBLE FOR ANY LOSS RESULTING FROM ANY USE OF ITS MATERIALS, PRODUCTS, SERVICES OR RECOMMENDATIONS DESCRIBED IN THIS DOCUMENT. Each user is responsible for making its own determination as to the suitability of Seller's materials, products, services or recommendations for the user's particular use through appropriate end-use and other testing and analysis. Nothing in any document or oral statement shall be deemed to alter or waive any provision of Seller's Standard Conditions of Sale or this Disclaimer, unless it is specifically agreed to in a writing signed by Seller. Statements by Seller concerning a possible use of any material, product, service or design do not, are not intended to, and should not be construed to grant any license under any patent or other intellectual property right of Seller or as a recommendation for the use of any material, product, service or design in a manner that infringes any patent or other intellectual property right.

SABIC and brands marked with ™ are trademarks of SABIC or its subsidiaries or affiliates.

† Any brands, products or services of other companies referenced in this document are the trademarks, service marks and/or trade names of their respective holders.

No. 19

Fixed Bed Reactor Fouling

Rentsch, C., Micromidas Inc

Fixed bed reactors are a workhorse in the chemical and petroleum refining industry. A reactor is often expected to run uninterrupted for multiple years between planned shutdowns. Chemical engineers are expected to skillfully monitor and manage the forecasted remaining catalyst life, as unplanned outages are significantly more costly than scheduled maintenance shutdowns.

Catalyst poisoning may occur when a hydrocarbon reactor receives feed material contaminated with non-hydrocarbon materials, such as metal ions. In addition to poisoning, the catalyst bed is subject to mechanical obstruction from solids in the feed. Catalyst extrudates are often very small to reduce mass transfer resistance, and a deep bed of fine catalyst extrudates is an efficient particulate capture system. Particulate fouling will be evidenced by increasing reactor differential pressure. Examples of undesired feed particles include carbonaceous materials, polymeric material, pipe scale, and inorganic suspended matter.

Either loss of catalyst activity or rising differential pressure may necessitate reactor throughput reduction or an expensive unplanned shutdown to replace the catalyst. In the event of a shutdown it is the job of the chemical engineer to adjust the catalyst loading plan to maximize the cycle life of the unit. Removing a portion of the catalyst and replacing it with grading material will increase tolerance to feed particulates but at the expense of fewer active catalyst sites and a higher deactivation rate. Increasing the amount of catalyst loaded either by dense-loading or by removing a portion of the grading layer volume may increase tolerance to poisons and reduce deactivation rates, but at the expense of reduced tolerance to feed particulate contamination.

In this engineer's experience, there is a third reactor fouling mode that is unusual and unexpected by most: particulate plugging of the bottom bed of a multi-bed reactor while the upper bed stays clean and free of particulate contamination, despite both beds containing the same size, type, and density catalyst. This failure mode will be discussed along with potential root causes.

PDS Process Development Symposium

AIChE
The Global Home of Chemical Engineers

The nub of successful mixing process research and scale-up

Victor A. Atiemo-Obeng, PhD, FAIChE
Retiree from The Dow Chemical Company

PROCESS DEVELOPMENT SYMPOSIUM, JUNE 2-4, 2015, HOUSTON, TX

PDS Process Development Symposium

AIChE
The Global Home of Chemical Engineers

Survey

- Course in college on mixing as one of Unit Operations?
- Lecture in college on mixing concepts or phenomena?
- Course outside college on mixing as one of Unit Operations?
- Lecture outside college on mixing concepts or phenomena?

6/16/2015

PROCESS DEVELOPMENT SYMPOSIUM, JUNE 2-4, 2015, HOUSTON, TX

2

AIChE
The Global Home of Chemical Engineers

Mixing - The Forgotten Unit Operation

"Although mixing equipment appears in nearly every process plant, mixing as a discipline is not routinely taught to undergraduates …"

SUZANNE M KRESTA (2011) CANADIAN J. OF CHEMICAL ENGINEERING, VOL. 89, OCTOBER 2011 p 959

6/16/2015

PROCESS DEVELOPMENT SYMPOSIUM, JUNE 2-4, 2015, HOUSTON, TX

3

AIChE
The Global Home of Chemical Engineers

Index Flowsheet - Chemical & Allied Processing Industry

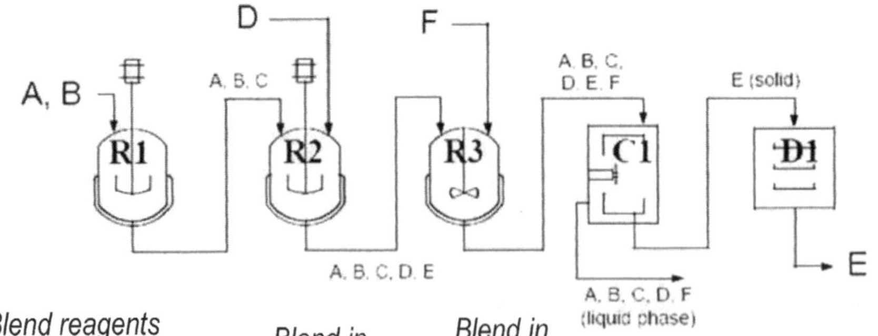

Blend reagents A and B Heat and React to form C Transfer to R2

Blend in reagent D React with C to form E

Blend in reagent F Cool and crystallize E

Separate E by centrifuge

Dry E

Effective mixing is required for success

6/16/2015

PROCESS DEVELOPMENT SYMPOSIUM, JUNE 2-4, 2015, HOUSTON, TX

4

AIChE
The Global Home of Chemical Engineers

Ideal reactors in Reaction Engineering

- uniform concentration and other properties
- instantaneous response to any process change

6/16/2015

AIChE
The Global Home of Chemical Engineers

Power correlation for agitated/stirred tank

$$Np \equiv \frac{P}{\rho N^3 D^5} = \Phi(\frac{(ND)D}{\upsilon}, \frac{N^2 D}{g}, Nt)$$

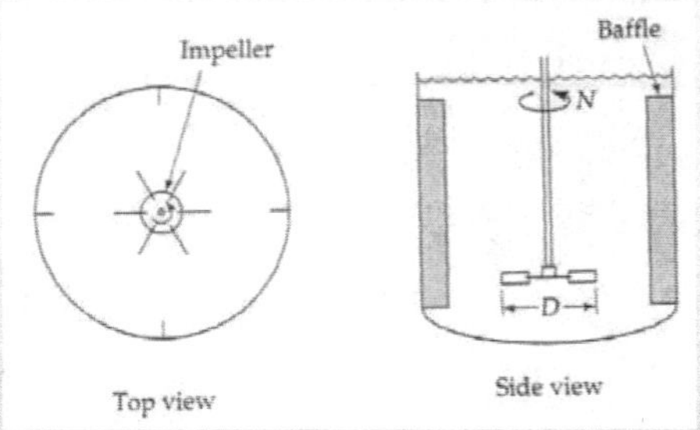

Np, Power Number [-]
P, Power input [W]
N, Impeller rotation [s-1]
D, Impeller diameter [m]
t, Time from start [s]
g, Gravitational acceleration [m2/s]
ρ, Density [Kg/m3]
ν, Kinematic viscosity [m2/s]

Bird, Stewart and Lightfoot, Transport Phenomena, Wiley 1960, p. 205

6/16/2015

PDS Process Development Symposium

AIChE The Global Home of Chemical Engineers

IMPORTANT LESSONS

Process failure from inadequate mixing is costly

- lower reaction yields, more by-products
- longer reaction/process times
- unacceptable product properties
- higher costs for purifications, waste handling, etc.
- higher production costs for product rework, etc.
- less safe operations

Early and careful assessment of effect of mixing on process is imperative!

PROCESS DEVELOPMENT SYMPOSIUM, JUNE 2-4, 2015, HOUSTON, TX

6/16/2015

7

PDS Process Development Symposium

AIChE The Global Home of Chemical Engineers

Describing mixing processes

Consider these cases:

A. Prepare a drink: Rum and Coke
B. Sweeten tea/coffee with honey or granular sugar
C. Make salad dressing: Oil and vinegar
D. Make bread dough: flour and water

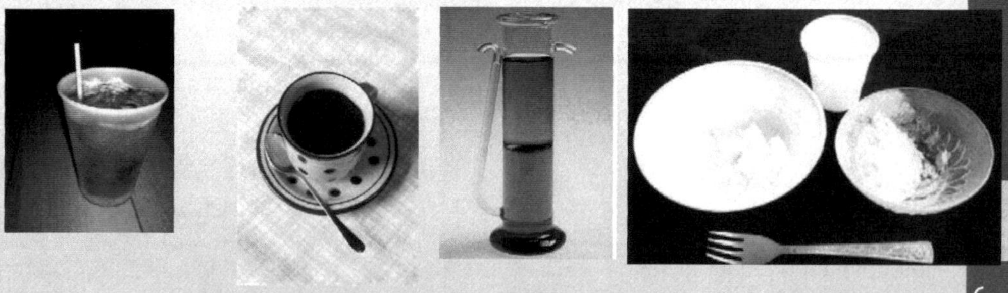

PROCESS DEVELOPMENT SYMPOSIUM, JUNE 2-4, 2015, HOUSTON, TX

6/16/2015

8

Describing mixing processes

A. Rum and Coke
B. Sweeten tea/coffee
C. Make salad dressing
D. Make bread dough

Compare based on:

- Desired process outcomes
- Phases involved
- Phenomena occurring
- Difficulty/challenge
- Equipment used

Chemical Process Development & Scale-up

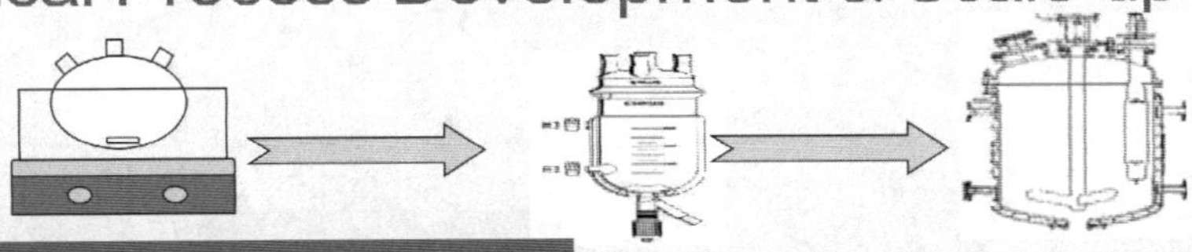

Physical properties

- Phases
- Density
- Viscosity & rheology

Chemistry

- Chemical equilibria
- Chemical kinetics

Transport phenomena

- Hydrodynamics/Mixing
- Mass transfer
- Heat transfer

PDS Process Development Symposium

AIChE The Global Home of Chemical Engineers

Stirred Tank Hydrodynamics:
Laminar or Turbulent flow?

Characterized by Reynolds Number, Re_I

Depends on - $Re_I = \rho(ND)D/\mu$

- fluid properties: density, ρ; viscosity, μ
- size or scale of equipment: impeller diameter, D

Consequences -

- Mixing mechanisms are different
- Equipment requirements differ

PROCESS DEVELOPMENT SYMPOSIUM, JUNE 2-4, 2015, HOUSTON, TX

6/16/2015

11

PDS Process Development Symposium

AIChE The Global Home of Chemical Engineers

Cases for consideration

- ***Solids suspension – effect of impeller diameter at equal power input on hydrodynamics***
- ***Multiphase Reactions - Effect of impeller rpm & reagent feed rate on process yield***
- ***Scaleup of Cooling crystallization - Effect of nucleation*** *temperature / degree of supersaturation*

PROCESS DEVELOPMENT SYMPOSIUM, JUNE 2-4, 2015, HOUSTON, TX

6/16/2015

12

Hydrodynamics of Solid-liquid Mixing

AIChE
The Global Home of Chemical Engineers

An eye opening experiment

$$P = Po\rho N^3 D^5 \qquad Tq = P/2\pi N$$

Relative P=1	Relative P=1	Relative P=1
Relative Tq=0.3	Relative Tq=1	Relative Tq=3

Courtesy of Lightnin

Hydrodynamics of Solid-liquid Mixing: Levels of suspension

AIChE
The Global Home of Chemical Engineers

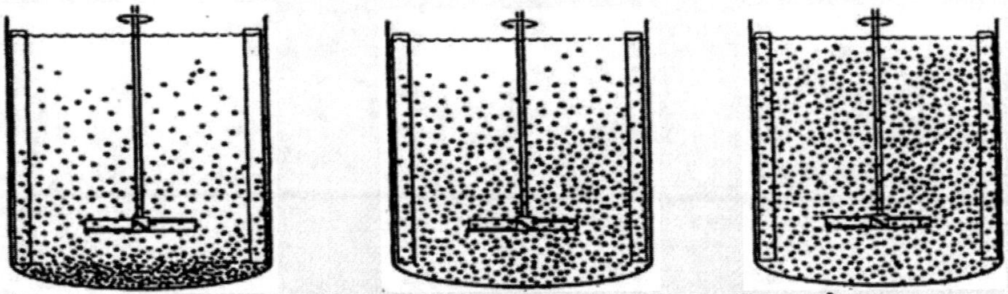

Increasing impeller rotational speed, N

Partial Suspension

- some solids rest on bottom of tank for short periods
- acceptable for dissolution of very soluble solids

Complete Off-Bottom/ "Just Suspension"

- no solids rest on bottom from more than a few seconds
- minimum condition for most applications

Uniform

- solids uniformly distributed
- required for crystallization, slurry feeds, etc.

AIChE
The Global Home of Chemical Engineers

Process Development & Scale-up: Role of Hydrodynamics

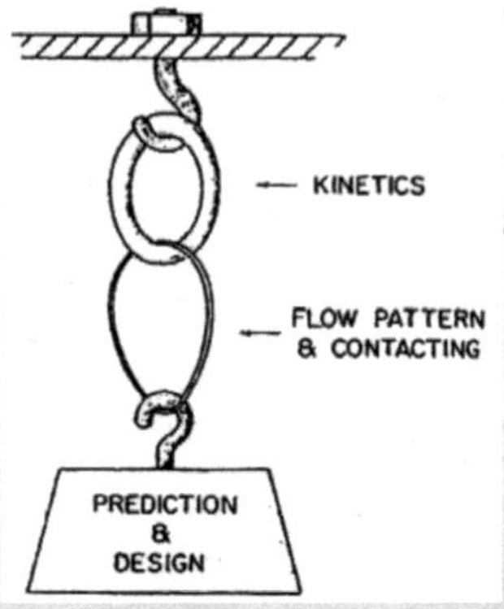

- Kinetics and hydrodynamics required for reactor scale-up & design
- "Often, the lack of knowledge of the **expected flow pattern in the reactor is the** main cause of uncertainty in the design of reactors, not the kinetics."

Levenspiel (1999) Ind. Eng. Chem. Res. 38, 4140

Effective stirring producing desired hydrodynamics is required for success

6/16/2015

AIChE
The Global Home of Chemical Engineers

Multiphase Reactions - Chloroacetylation: Effect of impeller rpm & reagent feed rate

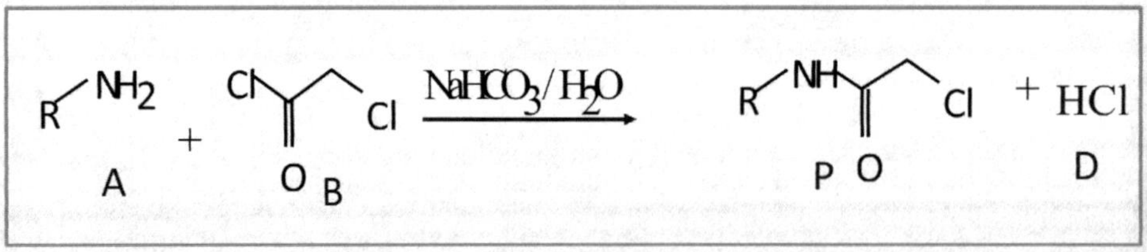

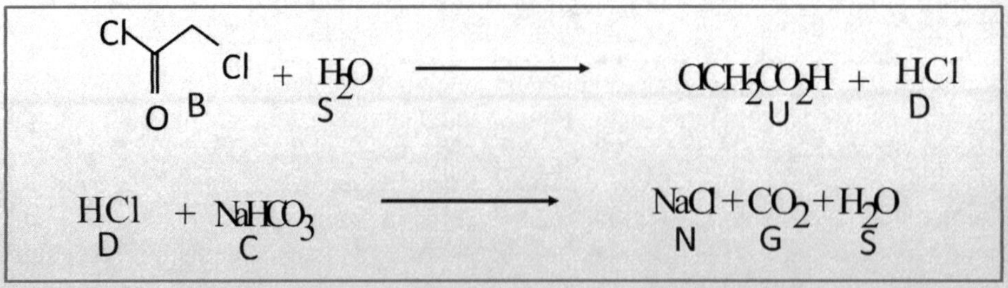

6/16/2015

PDS Process Development Symposium

AIChE The Global Home of Chemical Engineers

Multiphase Reactions - Chloroacetylation: Effect of impeller rpm & reagent feed rate

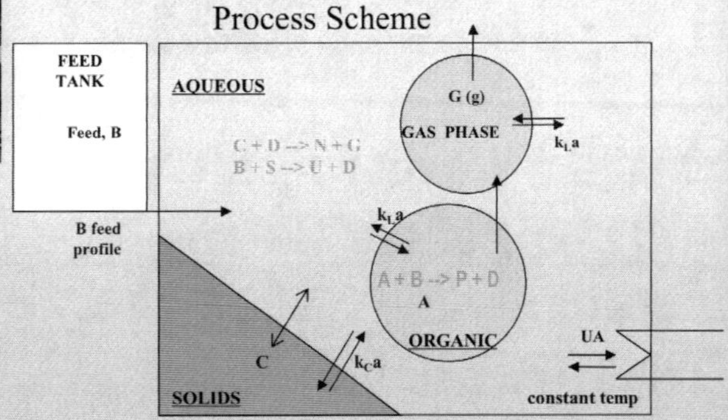

PROCESS DEVELOPMENT SYMPOSIUM, JUNE 2-4, 2015, HOUSTON, TX

6/16/2015

17

PDS Process Development Symposium

AIChE The Global Home of Chemical Engineers

Multiphase Reactions - Chloroacetylation: Effect of impeller rpm & reagent feed rate

Mixing Option	RPM	Feed Time	pH	Yield
2 PBT	300	3.7 hours	3.2	70%
2 PBT	400	3.5 hours	4.5	87%
2 PBT	400	5.7 hours	6.1	~90%

Mixing Option	RPM	Feed Time	pH	Yield
3 PBT	300	3.5 hours	5.3	87%
3 PBT	400	3.5 hours	6.6	91%
3 PBT	400-750	3.5 hours	6.6	94%

PROCESS DEVELOPMENT SYMPOSIUM, JUNE 2-4, 2015, HOUSTON, TX

6/16/2015

18

Scaleup of Cooling crystallization of Chloroketone

AIChE
The Global Home of Chemical Engineers

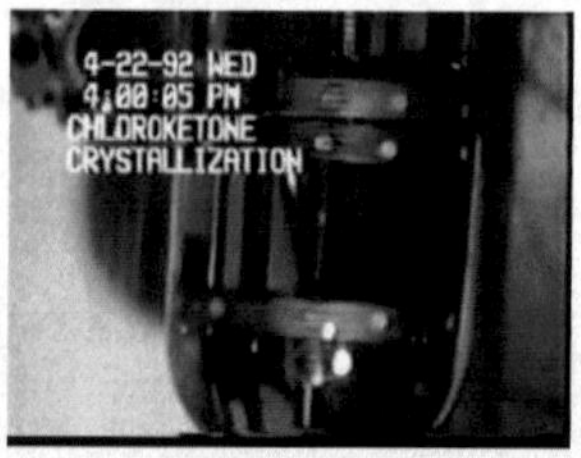

Desired process result

- Prevent encrustation on vessel wall
- Higher crystal yield
- Shorter batch time

PROCESS DEVELOPMENT SYMPOSIUM, JUNE 2-4, 2015, HOUSTON, TX

6/16/2015

19

Scaleup of Cooling crystallization of Chloroketone

AIChE
The Global Home of Chemical Engineers

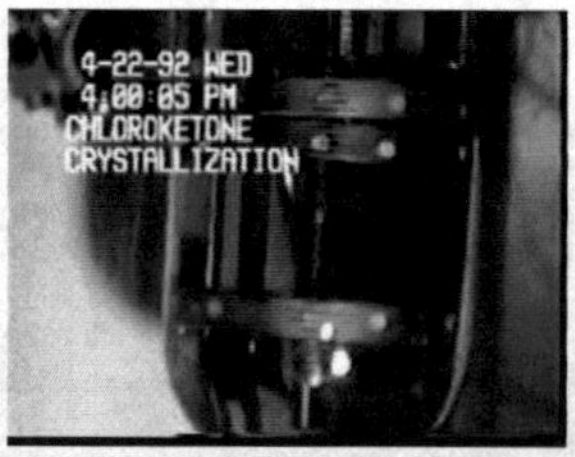

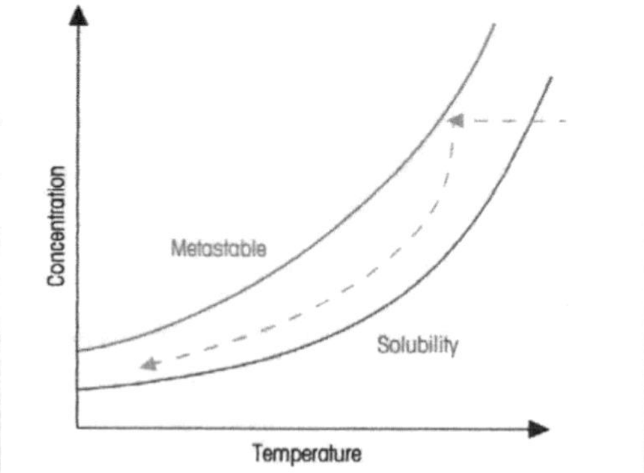

PROCESS DEVELOPMENT SYMPOSIUM, JUNE 2-4, 2015, HOUSTON, TX

6/16/2015

20

PDS Process Development Symposium

Scaleup of Cooling crystallization of Chloroketone

AIChE The Global Home of Chemical Engineers

PBT with baffles

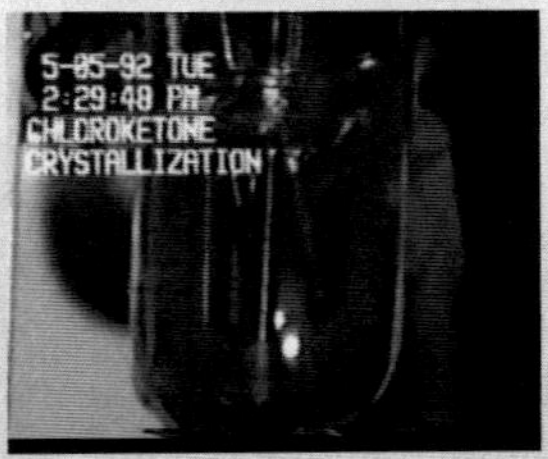

Anchor

Explore

- Impeller types
- Jacket temperature
- Nucleation temperature
- Seeding or not

PDS Process Development Symposium

Scaleup of Cooling crystallization of Chloroketone

AIChE The Global Home of Chemical Engineers

Run	Impeller	Jkt T (C)	Seeding	Nucleation T (C)	Encrustation
1	PBT/baffles	24	NO	25.3	YES
2A	Anchor	24	YES	30.0	NO
2B	Anchor	24	NO	31.6	NO
2C	Anchor	-10	NO	-2	SOLID BLOCK
2D	Anchor	10	NO	26	SEVERE
2E	Anchor	24	YES	25	YES
2F	Anchor	24	NO	27.7	YES

PDS Process Development Symposium

AIChE The Global Home of Chemical Engineers

Scaleup of Cooling crystallization of Chloroketone

6/16/2015

Solution

- System with few internal surfaces
- Monitor for nucleation
- Induce nucleation near 31°C

PROCESS DEVELOPMENT SYMPOSIUM, JUNE 2-4, 2015, HOUSTON, TX

23

PDS Process Development Symposium

AIChE The Global Home of Chemical Engineers

Summary

6/16/2015

✓ Focus on achieving desired process result
- ➢ Link "effective mixing" to achieving desired process result

✓ Describe/define effective mixing in terms of
- ➢ required hydrodynamics and relevant physicochemical phenomena necessary for process success
- ➢ confirm with basic calculations, experiments and/or modeling
- ➢ Recommend/select equipment and/or operating conditions to achieve desired process result

✓ Ascertain reliability of mechanical design

PROCESS DEVELOPMENT SYMPOSIUM, JUNE 2-4, 2015, HOUSTON, TX

24

IMPORTANT LESSON

AIChE
The Global Home of Chemical Engineers

"Fundamental knowledge must be coupled with practical insight and engineering judgment to solve problems associated with real industrial applications"

Leng & Calabrese (2004) *in Handbook of Industrial Mixing: Science and Practice,* Eds. E.L. Paul, V.A. Atiemo-Obeng and S.M. Kresta, Wiley & Sons.

Resources & Tools

AIChE
The Global Home of Chemical Engineers

http://www.1.mixing.net/

FMP Celebrating 30 years

Fluid Mixing Processes (FMP)
An Industrial Consortium

VisiMix
software for chemical engineering

TURBULENT® 2Kx
Low viscosity liquids and multi-phase systems

LAMINAR® 2Kx
Highly viscous and non-Newtonian media

http://www.visimix.com/

PDS Process Development Symposium

AIChE
The Global Home of Chemical Engineers

6/16/2015

Thank You!

PROCESS DEVELOPMENT SYMPOSIUM, JUNE 2-4, 2015, HOUSTON, TX

27

Chemical Processes

Siddall, J., Cabot Corporation

Chemical engineers are well equipped to analyze an existing or proposed manufacturing process to solve problems and to make sound decisions. Developing a new process uses these analysis skills as well as some new ones. This talk is about the process of process development, how it works, and what we achieve along the way. We don't make chemicals; we only control conditions that chemicals experience. Then what we want to happen occurs spontaneously. To do this, we need to know and exploit the behavior of the materials. We will also discuss the underlying drivers and realities of large processes and how these impact our work. Process development is especially fun and rewarding role for chemical engineers.

Estimating Physical Properties for Process Development

Dr. Kevin G. Joback
Molecular Knowledge Systems
http://www.molecularknowledge.com

Molecular Knowledge Systems

- ❑ **Located in Bedford, New Hampshire**
- ❑ **Company Started in 1989**
- ❑ **Computer Software**
- ❑ **Consulting**

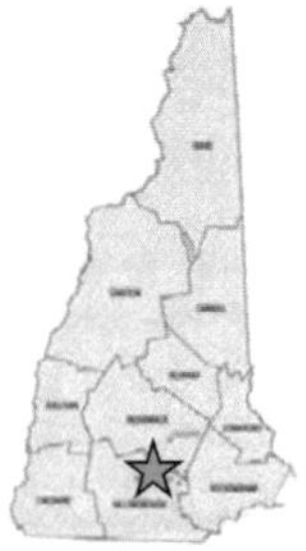

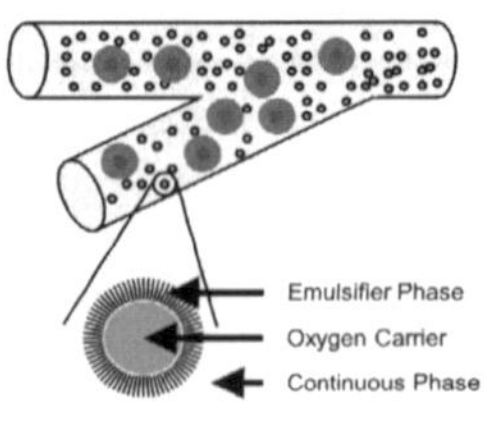

2000-11-15-01

Goal: Think about Physical Properties

Physical properties are often investigated only after many process decisions have been made.

- ❑ **Physical properties define the process**
- ❑ **Calculate properties when you can**
- ❑ **Understand properties always**

Subgoal: 1) Group Contribution Techniques; 2) Equation Oriented Techniques; 3) Mixture Techniques

2000-11-15-01

CO_2 Solvent Absorption

Solvent Selection

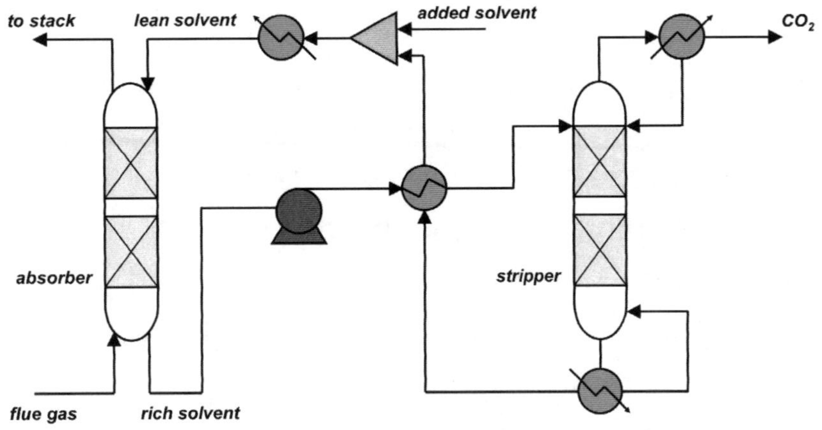

2000-11-15-01

Gas Absorption: Solvent

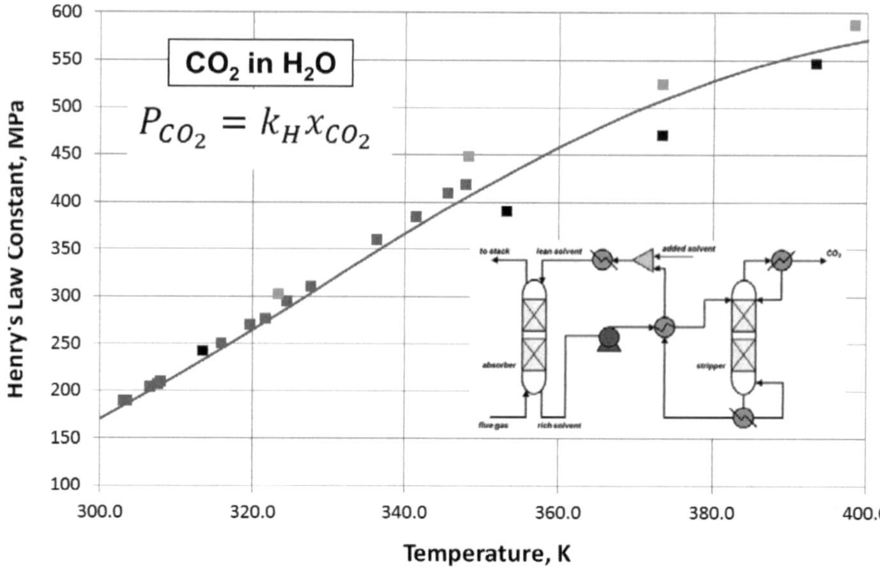

Solvent Properties

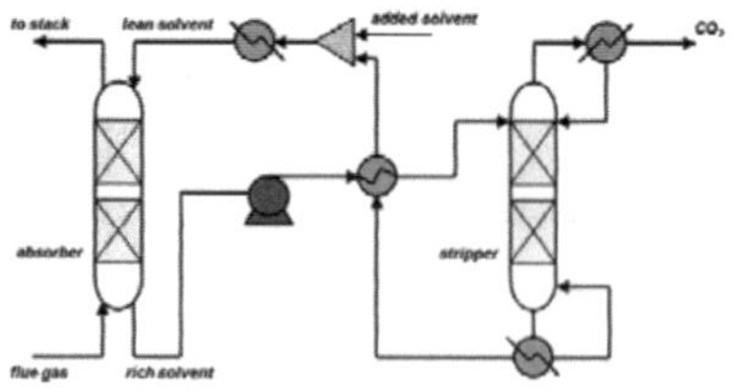

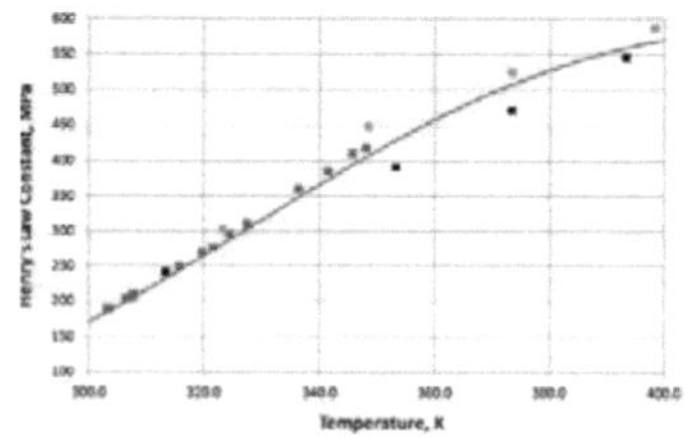

- ❑ **Change absorption with temperature:** k_H
- ❑ **Good heat and mass transfer:** η, λ, σ, D_{ab}
- ❑ **Liquid at common conditions:** T_m, T_b
- ❑ **Should require low energy:** C_p, ΔH
- ❑ **Safe to use:** T_f, LD_{50}, LC_{50}
- ❑ **Thermally stable:** $T_{d,0}$

Common Solvents

Monoethanolamine

$OH{-}CH_2{-}CH_2{-}NH_2$

Diethanolamine

$OH{-}CH_2{-}CH_2{-}NH{-}CH_2{-}CH_2{-}OH$

2-Amino-2-methylpropanol

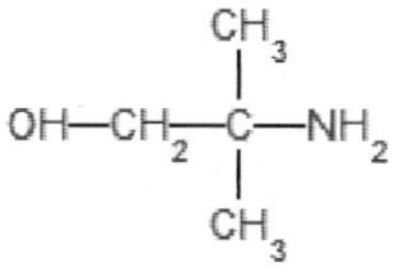

2000-11-15-01

New Solvents

$-CH_3$ $-CH_2-$ $>CH-$ $>C<$ $-OH$ $-NH_2$ $-NH-$ $>N-$

All Combinations

(2 0 0 0 0 0 0 0)
(2 1 0 0 0 0 0 0)
(2 2 0 0 0 0 0 0)
(2 3 0 0 0 0 0 0)
(2 4 0 0 0 0 0 0)
•••
(2 0 0 1 1 0 0 0)

$$\frac{b}{2} = n - 1 + r$$

b = number of bonds
n = number of groups
r = number of rings

$$\frac{7}{2} = 4 - 1 + 0 \quad \times$$

2000-11-15-01

New Solvents

-CH_3 -CH_2- >CH- >C< -OH -NH_2 -NH- >N-

All Combinations
(2 0 0 0 0 0 0 0)
(2 1 0 0 0 0 0 0)
(2 2 0 0 0 0 0 0)
(2 3 0 0 0 0 0 0)
(2 4 0 0 0 0 0 0)
•••
(2 2 2 0 2 0 1 0)

$$\frac{b}{2} = n - 1 + r$$

b = number of bonds
n = number of groups
r = number of rings

$$\frac{16}{2} = 9 - 1 + 0 \;\checkmark$$

2000-11-15-01

Group Contribution Techniques

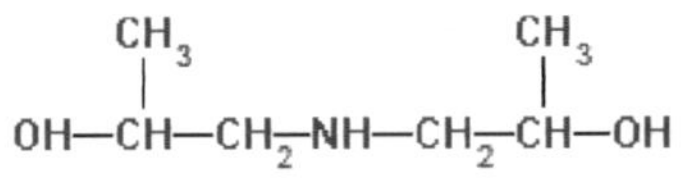

Group	n_i	$\Delta_{i,Tm}$
-CH3	2	-5.10
-CH2-	2	11.27
>CH-	2	12.64
-OH	2	44.45
-NH-	1	52.66

Model:
$$Tm = 122 + \Sigma n_i \Delta_i$$

Tm (est) = 301.18 K
Tm (lit) = 315.15 K

2000-11-15-01

Boiling Point, Estimation Errors

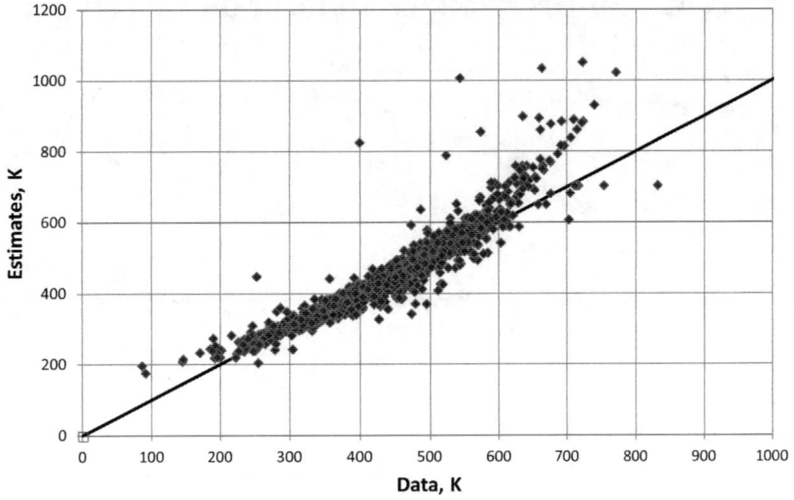

Estimates generated using Joback's method

2000-11-15-01

Boiling Point, Estimation Errors

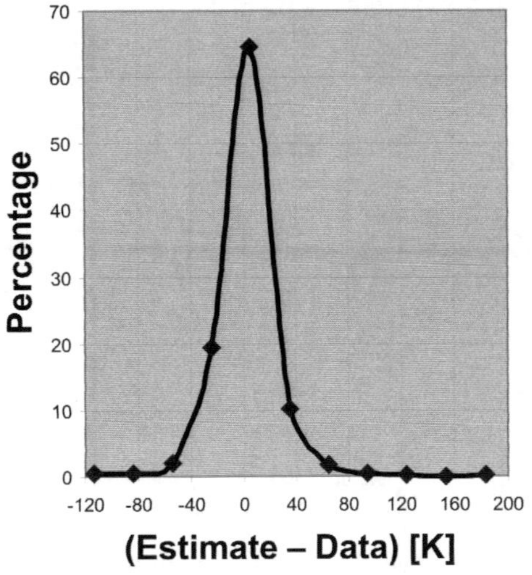

Estimates generated using Joback's method

Statistics

Observations	559
Avg Error	0.97 K
Avg Abs Err	15.1 K
Avg % Error	4.8 %
Max Error	197.4 K

Outliers, Errors

N-Methylformamide	-128.5 K
Acetamide	-122.6 K
Fluorine	113.1 K
Cyanogen	197.4 K

2000-11-15-01

Group Contribution Techniques

Normal Paraffins Homologous Series

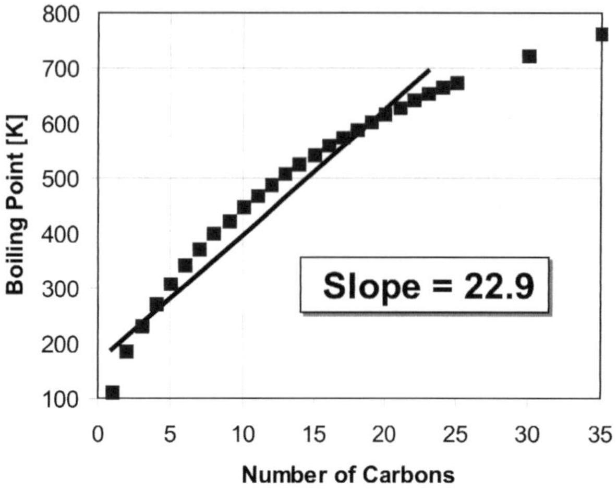

$CH_3 - [CH_2]_n - CH_3$

Contribution

-CH2- = 22.9

2000-11-15-01

Group Representation

Groups: -CH3 -CH2- >CH- >C<

Octane Isomer	Tm, lit [K]	Tm, est [K]
	163.0	164.4
	152.0	164.4
	151.0	164.4
	172.0	166.8
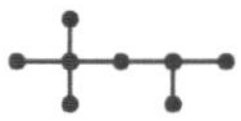	165.8	166.8
	160.5	166.8

Estimates generated using Joback's method

2000-11-15-01

CO_2 Solvent Absorption

Solvent Selection

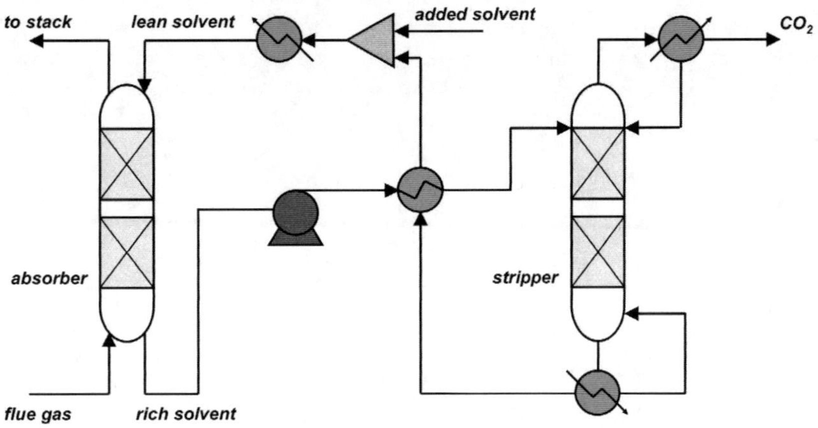

2000-11-15-01

Equation Oriented Techniques

Vapor Pressure: Ambrose + Walton Method

$$\ln\left(\frac{P_{vp}}{P_c}\right) = f^{(0)} + \omega f^{(1)} + \omega^2 f^{(2)}$$

$$f^{(0)} = \frac{-5.97616\tau + 1.29874\tau^{1.5} - 0.60394\tau^{2.5} - 1.06841\tau^5}{T/T_c}$$

$$f^{(1)} = \frac{-5.03365\tau + 1.11505\tau^{1.5} - 5.41217\tau^{2.5} - 7.46628\tau^5}{T/T_c}$$

$$f^{(2)} = \frac{-0.64771\tau + 2.41539\tau^{1.5} - 4.26979\tau^{2.5} + 3.25259\tau^5}{T/T_c}$$

$$\tau = (1.0 - T/T_c)$$

2000-11-15-01

Equation Oriented Techniques

From Group Contribution Techniques:

$T_m = 235.5$ K $T_b = 409.4$ K $T_c = 575.9$ K $P_c = 4126.5$ kPa

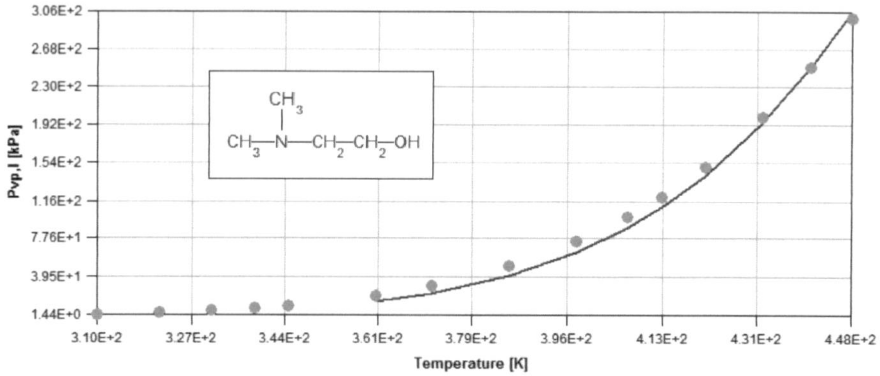

2000-11-15-01

Equation Oriented Techniques

Applicability Limits

$$\ln\left(P_{vp}\right) = A + \frac{B}{T}$$

2 Unknowns, 2 Points

T = Critical Temperature
P_{vp} = Critical Pressure

T = Boiling Point
P_{vp} = 1 atm

Heptane

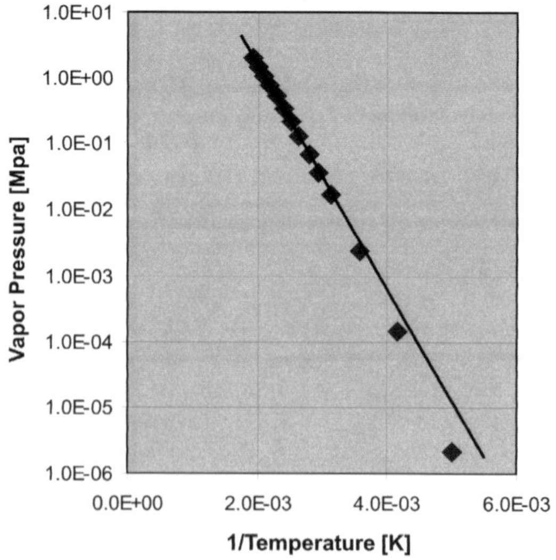

2000-11-15-01

Mixture Techniques

$$\ln(\eta_m) = x_1 \ln(\eta_1) + x_2 \ln(\eta_2) + x_1 x_2 G$$

Mixture Functions

- ❑ Pure Data Inputs
- ❑ Reduces to Pure
- ❑ Molecular Forces
- ❑ Characteristics

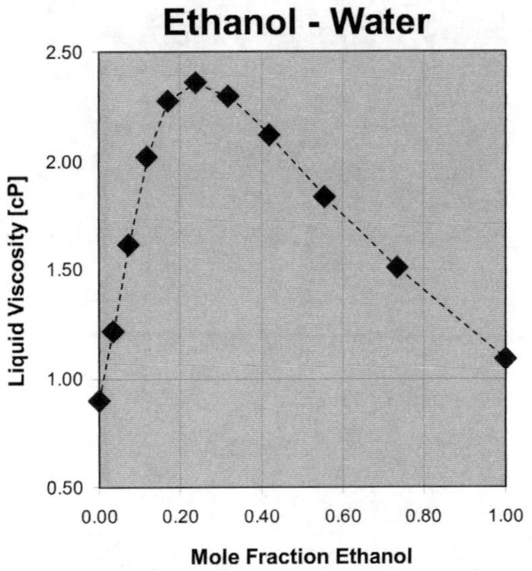

2000-11-15-01

Surface Tension – f(x)

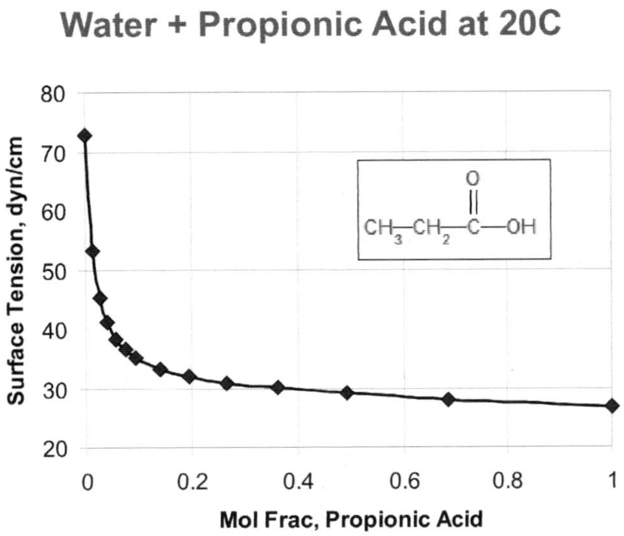

2000-11-15-01

Flash Points, Mixtures

Le Chatelier's Method

$$LFL_{mix} = \frac{1}{\sum \frac{y_i}{LFL_i}}$$

VLE Calculation

$$LFL_{mix} = \sum y_i$$

$$y_i = x_i \gamma_i P_i^{vp}(T_f)$$

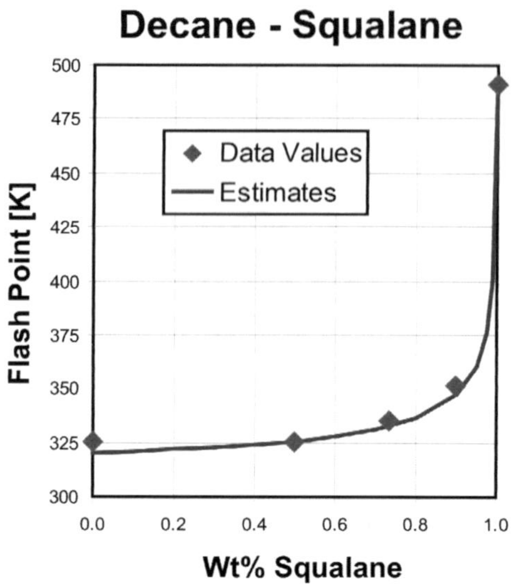

2000-11-15-01

Multiphase Viscosity

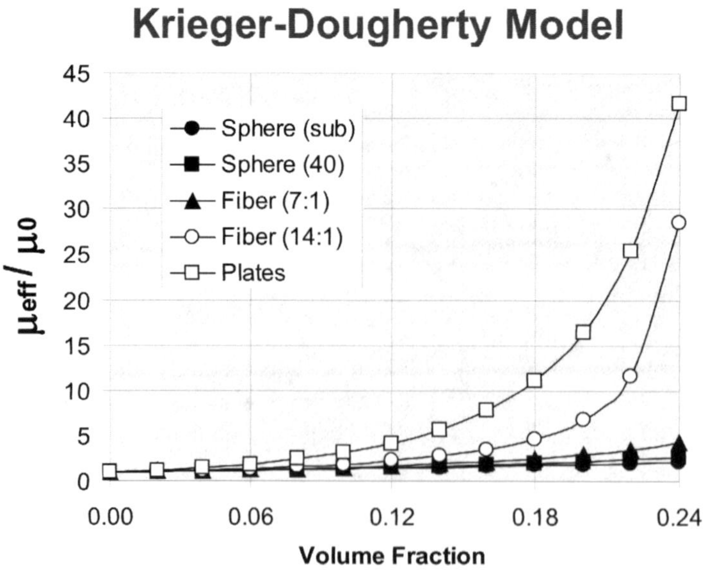

2000-11-15-01

Goal: Think about Physical Properties

Physical properties are often investigated only after many process decisions have been made.

- ❑ **Physical properties define the process**
- ❑ **Calculate properties when you can**
- ❑ **Understand properties always**

Subgoal: 1) Group Contribution Techniques; 2) Equation Oriented Techniques; 3) Mixture Techniques

2000-11-15-01

CO_2 Solvent Absorption

Solvent Selection

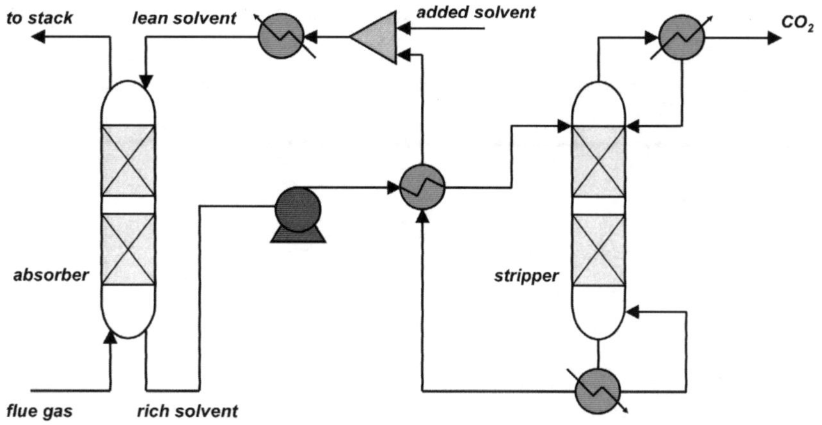

2000-11-15-01

Mass Transfer

Gas Phase, Liquid Phase, Wetted Area

$$k_G = 0.0338\left(\frac{D}{a_t d_p^2}\right)\left(\frac{\rho G}{a_t \mu}\right)^{0.8}\left(\frac{\mu}{\rho D}\right)^{0.33}$$

$$k_L = 0.0051\left(a_t d_p\right)^{0.4}\left(\frac{\mu g}{\rho}\right)^{0.33}\left(\frac{\rho L}{a_w \mu}\right)^{0.67}\left(\frac{\mu}{\rho D}\right)^{0.5}$$

$$\frac{a_w}{a_t} = 1 - exp\left\{-1.45\left(\frac{\sigma_c}{\sigma}\right)^{0.75}\left(\frac{\rho L}{a_t \mu}\right)^{0.1}\left(\frac{g}{L^2 a_t}\right)^{0.05}\left(\frac{\rho L^2}{a_t \sigma}\right)^{0.2}\right\}$$

L = liquid velocity
G = vapor velocity
ρ = solvent density
D = diffusion coefficient
σ_c = critical surface tension
σ = solvent surface tension
g = gravitational constant
d_p = packing diameter
μ = solvent viscosity
a_w = wetted area
a_t = total area

2000-11-15-01

Mass Transfer

Liquid Phase

$$k_L = C_L \frac{\mu^{0.16}}{\rho^{0.16} D^{0.5} a_w^{0.67}}$$

$$\frac{a_w}{a_t} = 1 - exp\left\{-C_A \frac{\rho^{0.3}}{\mu^{0.1}\sigma^{0.95}}\right\}$$

ρ = solvent density
D = diffusion coefficient
μ = solvent viscosity
σ = surface tension
a_w = wetted area

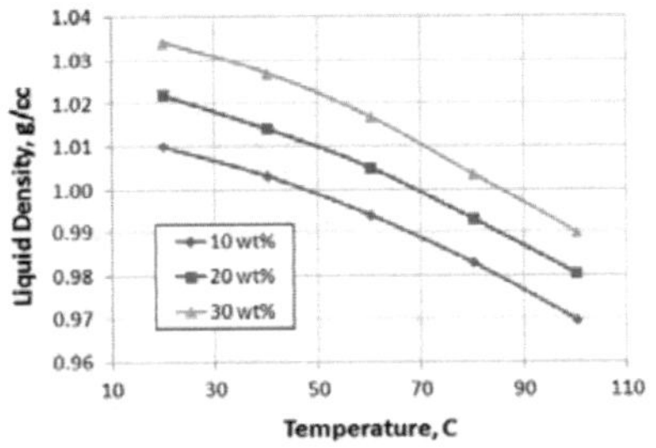

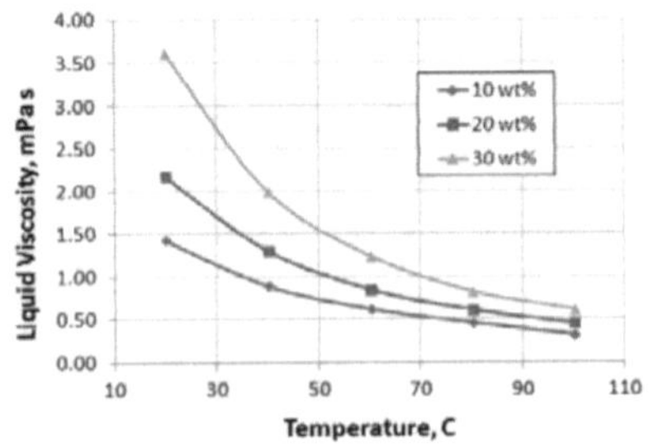

2000-11-15-01

Goal: Think about Physical Properties

Physical properties are often investigated only <u>after</u> many process decisions have been made.

- ❑ **Physical properties <u>define</u> the process**
- ❑ **<u>Calculate</u> properties when you can**
- ❑ **<u>Understand</u> properties always**

Subgoal: 1) Group Contribution Techniques; 2) Equation Oriented Techniques; 3) Mixture Techniques

2000-11-15-01

Chemical Products Too!

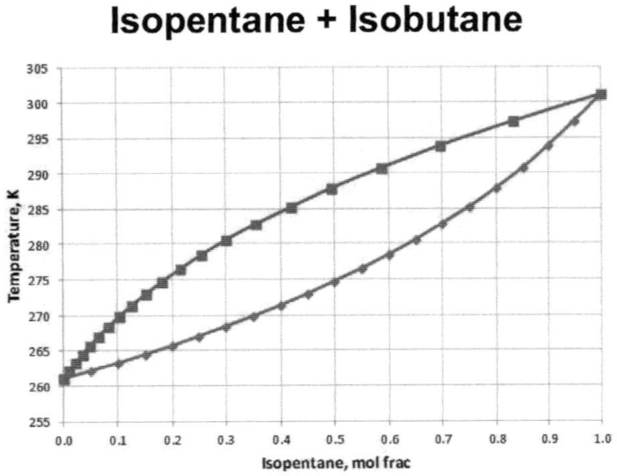

2000-11-15-01

Thank You !

Questions ?

2000-11-15-01

Molecular Knowledge Systems, Inc. *Page 29*

PDS Process Development Symposium

AIChE
The Global Home of Chemical Engineers

Computational Design of Catalytic Materials for Fuels & Chemicals Production using Activity Descriptors

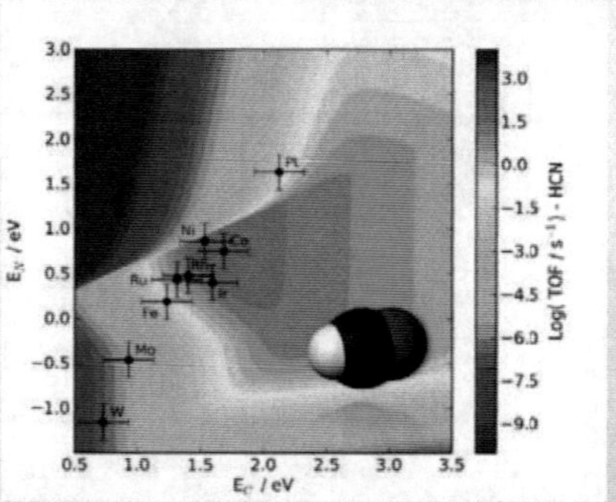

Lars C. Grabow
Chemical and Biomolecular Engineering
Chemistry
University of Houston
grabow@uh.edu

PROCESS DEVELOPMENT SYMPOSIUM, JUNE 2-4, 2015, HOUSTON, TX

PDS Process Development Symposium

AIChE
The Global Home of Chemical Engineers

Heterogeneous Catalysis: How much do we know?

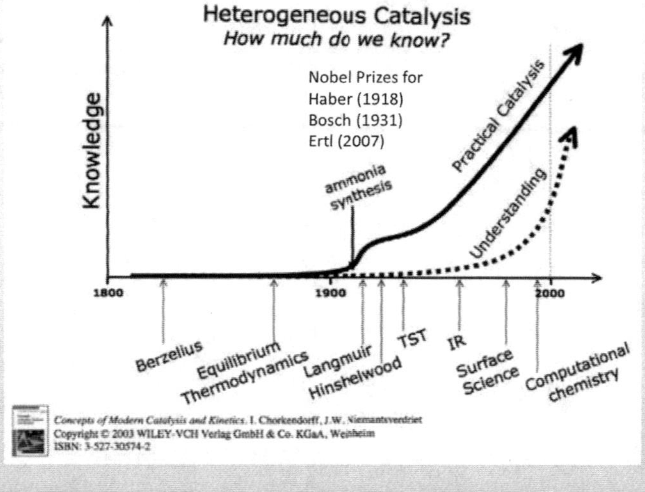

Concepts of Modern Catalysis and Kinetics. I. Chorkendorff, J.W. Niemantsverdriet
Copyright © 2003 WILEY-VCH Verlag GmbH & Co. KGaA, Weinheim
ISBN: 3-527-30574-2

2

PROCESS DEVELOPMENT SYMPOSIUM, JUNE 2-4, 2015, HOUSTON, TX

PDS Process Development Symposium

AIChE
The Global Home of Chemical Engineers

Kohn-Sham Density Functional Theory

- Born-Oppenheimer Approximation

$$E_{total} = E_{nucl}\left(\{R_\alpha\}\right) + E_{elec}\left(\{R_\alpha\}\right)$$

- **Theorem I:** The ground state properties of a many-electron system is uniquely determined by the electron density n(r) (3 variables).

$$E = E[n(\mathbf{r}), \{\mathbf{R}_\alpha\}] \quad n(\mathbf{r}) = \sum\nolimits_{i,occ} |\psi_i(\mathbf{r})|^2$$

- **Theorem II:** The total energy E[n] has a minimum for the ground state density $n_0(r)$.

Noble Prize in Chemistry
Walter Kohn (left) 1998

Hohenberg, Kohn, *Phys. Rev.* **136**, B862, **1964**
Kohn, Sham, *Phys. Rev.* **140**, A1133, **1965**

3

PROCESS DEVELOPMENT SYMPOSIUM, JUNE 2-4, 2015, HOUSTON, TX

PDS Process Development Symposium

AIChE
The Global Home of Chemical Engineers

The Bus Route

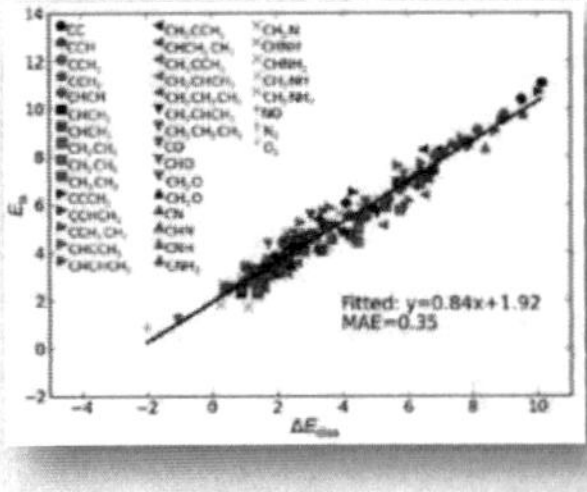

Identification of Reactivity Descriptors

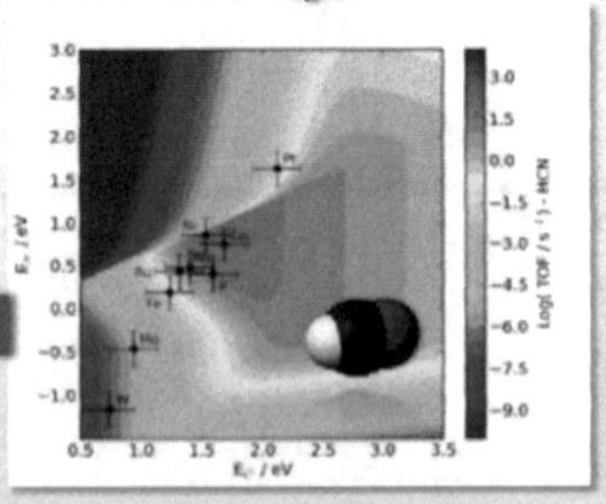

Descriptor-based discovery of metal alloys

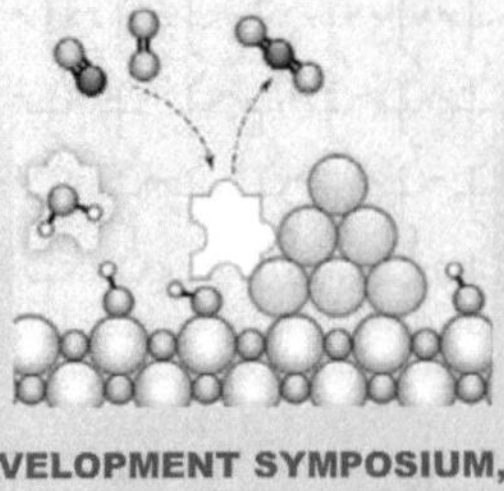

Description of metal/metal-oxide interfaces

4

PROCESS DEVELOPMENT SYMPOSIUM, JUNE 2-4, 2015, HOUSTON, TX

AIChE
The Global Home of Chemical Engineers

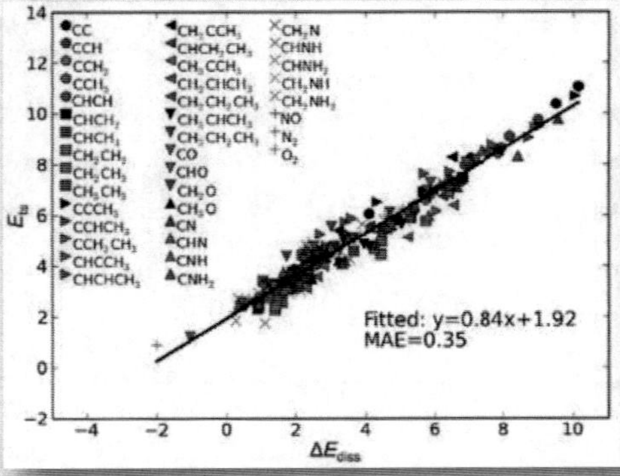

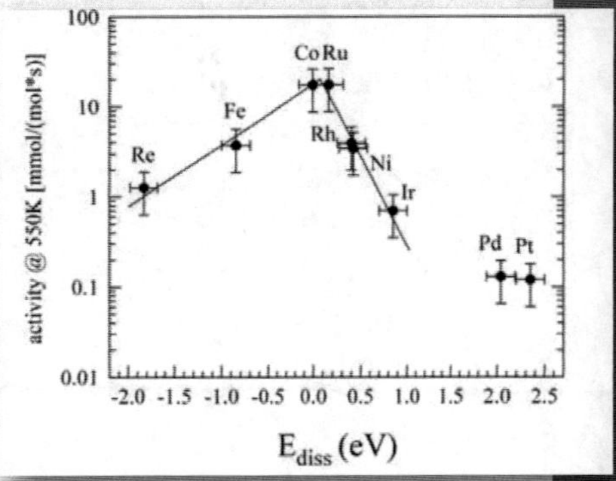

IDENTIFICATION OF REACTIVITY DESCRIPTORS

Wang *et al.*, *Catal. Lett.* **141**, 370, **2011**
Bligaard, Nørskov, Dahl, Matthiesen, Christensen, Sehested, J.Catal. **224**, 206, **2004**

AIChE
The Global Home of Chemical Engineers

Sabatier Volcano

Sabatier Principle: "The interactions between adsorbates and the catalyst should be neither too strong, nor too weak. They should be just right!"

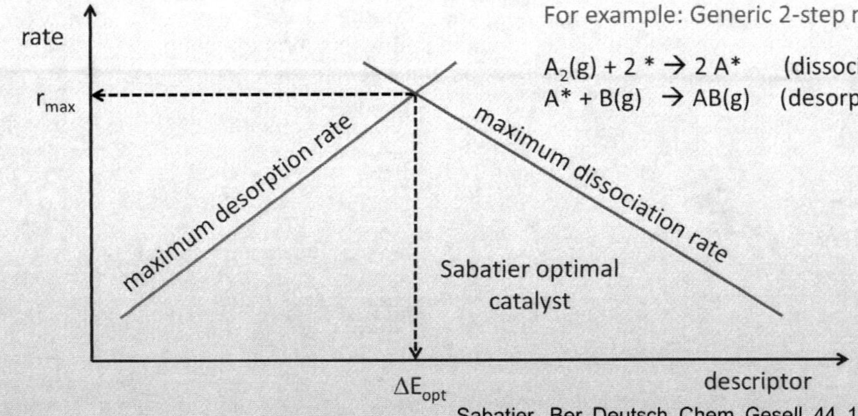

For example: Generic 2-step reaction

$A_2(g) + 2 * \rightarrow 2 A*$ (dissociation)
$A* + B(g) \rightarrow AB(g)$ (desorption)

Sabatier, Ber. Deutsch. Chem. Gesell. 44, 1984, **1911**

AIChE
The Global Home of Chemical Engineers

Linear Scaling Relations

CH_x adsorption energies | **NH_x adsorption energies**

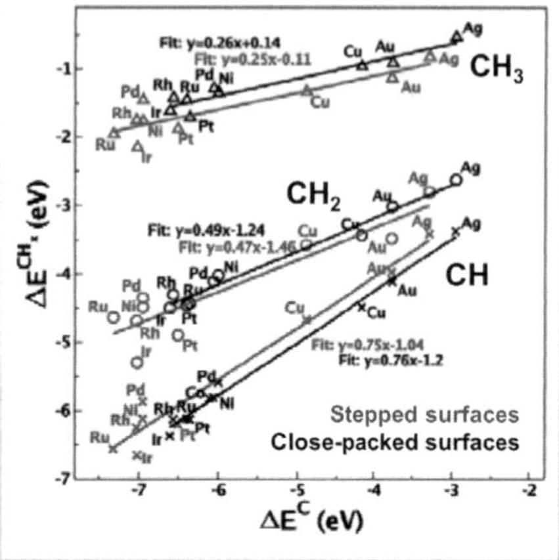

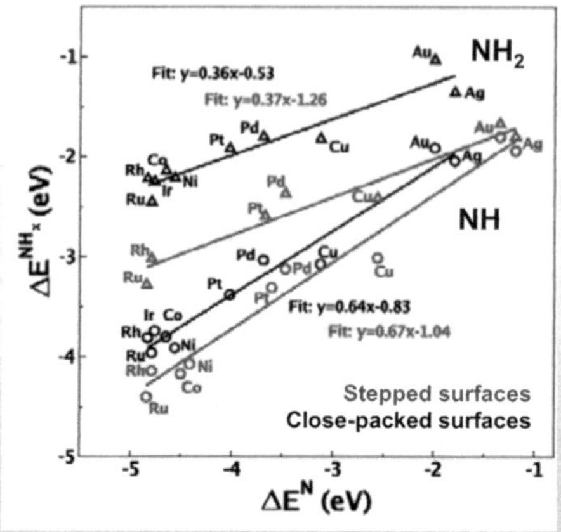

Abild-Pedersen *et al*, PRL **99,** 016105, **2007**

AIChE
The Global Home of Chemical Engineers

Brønsted-Evans-Polanyi (BEP) Relation

"Scaling laws for transition states"

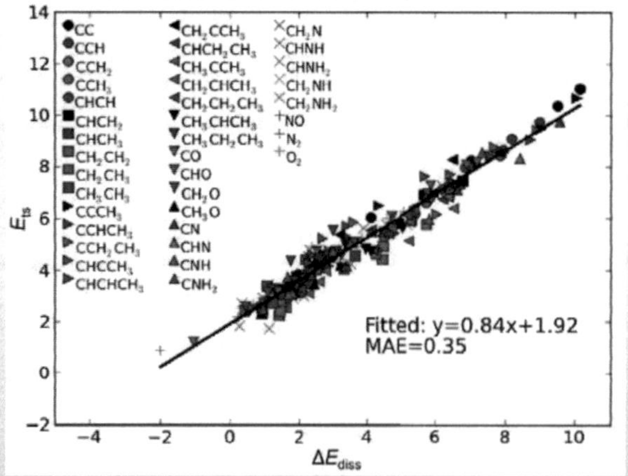

There exists a nearly linear (universal) relationship between E_a and the heat of dissociation ΔE_{diss}.

Several outliers exist, but a strong trend can easily be identified.

Wang *et al.*, *Catal. Lett.* **141**, 370, **2011**

AIChE

The Global Home of Chemical Engineers

Why Do Descriptors Work? The d-Band Model

Video Presentation in Original e-media

$$\Delta E = \Delta E_{sp} + \Delta E_{hyb-d}$$

Hammer, Nørskov, *Surf. Sci.* **343**, 211, **1995**
Hammer, Nørskov, *Adv. Catal.* **45**, 71, **2000**
Bligaard, Nørskov in *Chemical bonding at surfaces, Elsevier,* **2008**

9

PROCESS DEVELOPMENT SYMPOSIUM, JUNE 2-4, 2015, HOUSTON, TX

AIChE

The Global Home of Chemical Engineers

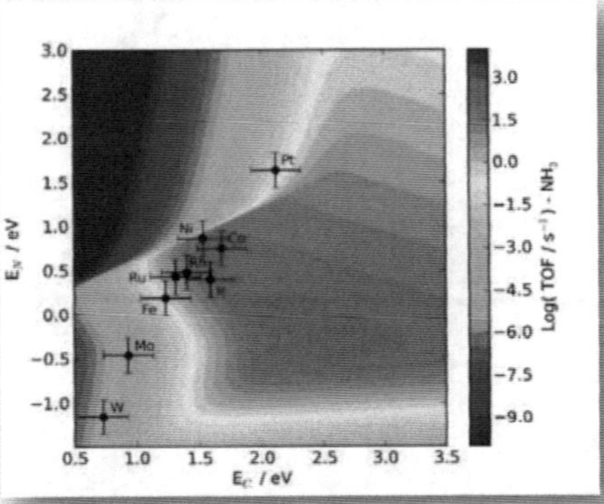

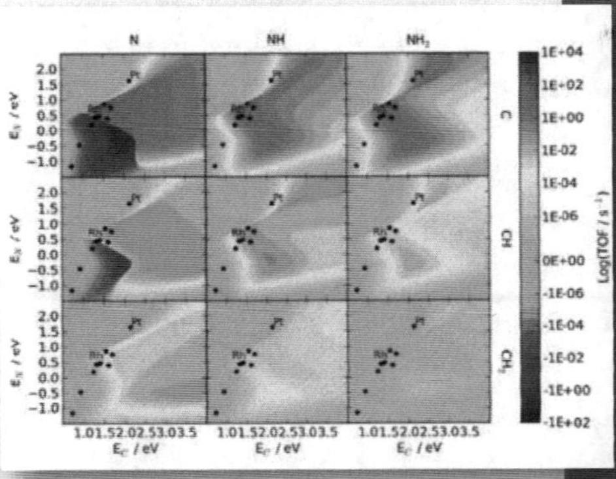

DESCRIPTOR-BASED DISCOVERY OF METAL ALLOYS: HCN SYNTHESIS

10

Grabow, Studt, Abild-Pedersen, Petzold, Kleis, Bligaard, Nørskov, *Angew. Chem. Int. Ed.* **50**, 4601, **2011**

PROCESS DEVELOPMENT SYMPOSIUM, JUNE 2-4, 2015, HOUSTON, TX

AIChE
The Global Home of Chemical Engineers

Hydrogen Cyanide: Synthesis and Applications

- HCN is a valuable precursor and versatile building block:
  - Methyl methacrylate (MMA), methionine, adiponitrile (nylon), sodium cyanide, chelating agents (EDTA, NTA), ...
- Large scale industrial production since 1940 in Germany and USA
- Typical HCN plant produces (5-30)$\times 10^3$ t/a
  - DuPont, Rohm and Haas/DOW, Novartis, Cyanco, ICI, Butachemie, Röhm, Elf-Atochem, Mitsubishi Gas.
- T = 1200-1500 K, atmospheric pressure, short contact times (ca. 1 s) over Pt-based catalysts.
- Energy consumption: 4 × 106 kJ/100 kg HCN.

Ullmann's Encyclopaedia of Industrial Chemistry, Release 2010, 7th Ed., Wiley-VCH, Weinheim, **2010**

PROCESS DEVELOPMENT SYMPOSIUM, JUNE 2-4, 2015, HOUSTON, TX

11

AIChE
The Global Home of Chemical Engineers

HCN Synthesis: A Selectivity Challenge

Thermodynamic Selectivity

$NH_3 + CH_4 \rightarrow HCN + 3\,H_2$
$2\,NH_3 \rightarrow N_2 + 3\,H_2$

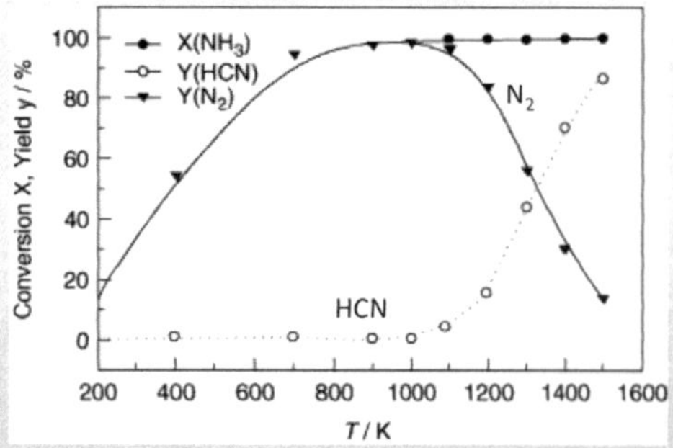

Typical outlet composition of the reaction gases as vol%

Compound	Andrussow process	BMA process
HCN	7.0	22.9
NH_3	1.7	2.5
CH_4	0.2	1.7
N_2	52.5	1.1
H_2	12.0	71.8
H_2O	23.0	—
CO	3.3	—
CO_2	0.3	—

"The more effectively a substance catalyzes the decomposition of ammonia, the less suitable it is for BMA synthesis." Koberstein (Degussa), 1953.

PROCESS DEVELOPMENT SYMPOSIUM, JUNE 2-4, 2015, HOUSTON, TX

12

PDS Process Development Symposium

AIChE
The Global Home of Chemical Engineers

Reaction Mechanism of the BMA

- C-monolayer reacts with NH_x fragments:
  - Hasenberg, Schmidt, *J. Catal.* 1985, **91**, 116; *J. Catal.* 1986, **97**, 156-168; *J. Catal.* 1987, **104**, 441-453.
- Coupling between atomic C and N:
  - Herceg, Trenary, *J. Am. Chem. Soc.* 2003, **125**, 15758-15759; *J. Phys. Chem. B* 2005, **109**, 17560-17566.
- NH_3 decomposition is rate limiting; coupling between C-N and HC-N, but other CH_x-NH_y mechanisms cannot be excluded:
  - Delagrange, Schuurman, *Catal. Today* 2007, **121**, 204-209
- Role of gas phase reactions:
  - Diefenbach et al., *J. Am. Chem. Soc.* 1999, **121**, 10614-10625
  - Horn et al., *PCCP* 2004, **6**, 4514-4521

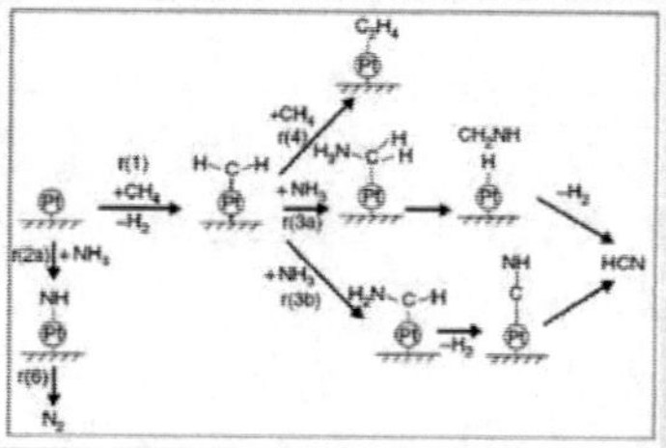

M. Diefenbach et al., *J. Am. Chem. Soc.* **121**, 10614, **1999**

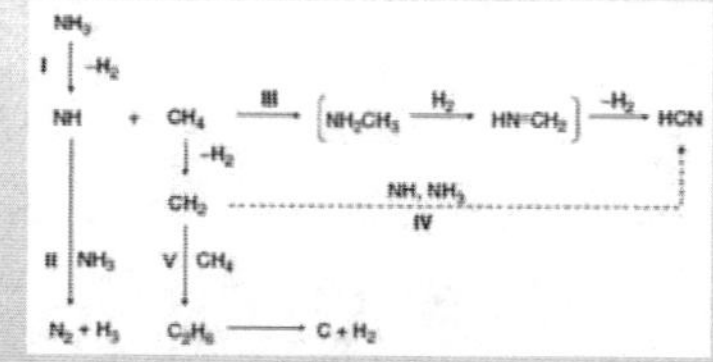

Pfeil, Hofmann, *Ber. Bunsen-Ges. Phys. Chem.* **67**, 229, **1963**

13

PROCESS DEVELOPMENT SYMPOSIUM, JUNE 2-4, 2015, HOUSTON, TX

PDS Process Development Symposium

AIChE
The Global Home of Chemical Engineers

HCN Synthesis Reactions

CH_4 decomposition	C-N coupling	NH_3 decomposition
$CH_4 + 2\,{*} \rightarrow CH_3{*} + H{*}$	$CH_2{*} + NH_2{*} \rightarrow H_2CNH_2{*} + {*}$	$NH_3 + 2\,{*} \rightarrow NH_2{*} + H{*}$
$CH_3{*} + {*} \rightarrow CH_2{*} + H{*}$	$CH_2{*} + NH{*} \rightarrow H_2CNH{*} + {*}$	$NH_2{*} + {*} \rightarrow NH{*} + H{*}$
$CH_2{*} + {*} \rightarrow CH{*} + H{*}$	$CH_2{*} + N{*} \rightarrow H_2CN{*} + {*}$	$NH{*} + {*} \rightarrow N{*} + H{*}$
$CH{*} + {*} \rightarrow C{*} + H{*}$	$CH{*} + NH_2{*} \rightarrow HCNH_2{*} + {*}$	$2\,N{*} \rightarrow N_2{*} + {*}$
$2\,H{*} \rightarrow H_2{*} + {*}$	$CH{*} + NH{*} \rightarrow HCNH{*} + {*}$	$N_2{*} \rightarrow N_2 + {*}$
$H_2{*} \rightarrow H_2 + {*}$	$CH{*} + N{*} \rightarrow HCN{*} + {*}$	
	$C{*} + NH_2{*} \rightarrow CNH_2{*} + {*}$	
	$C{*} + NH{*} \rightarrow CNH{*} + {*}$	
	$C{*} + N{*} \rightarrow CN{*} + {*}$	
	$CH_xNH_y{*} + (x+y-1){*} \rightarrow HCN{*} + (x+y-1)\,H{*}$	(fast)
	$HCN{*} \rightarrow HCN + {*}$	

14

PROCESS DEVELOPMENT SYMPOSIUM, JUNE 2-4, 2015, HOUSTON, TX

AIChE

The Global Home of Chemical Engineers

Microkinetic Model

- Transition state theory for each elementary step

$$k_i^+ = A \cdot e^{(-E_a)/k_B T} \qquad k_i^- = \frac{k_i^+}{K_i} \qquad K_i = e^{-\Delta G_i / k_B T} = e^{(-\Delta H_i + T\Delta S_i)/k_B T}$$

- Reaction network with 27 elementary steps
  - 27 values for ΔH, ΔS, A, E_a = 108 parameters
- This model too complex for screening purposes and simplifications are necessary!
- Solution: Find small set of descriptors that allows us to investigate trends
  - Scaling relations (stability of intermediates)
  - Brønsted-Evans-Polanyi (BEP) relations (stability of transitions states, activation barriers)

15

AIChE

The Global Home of Chemical Engineers

Putting the Puzzle Pieces Together

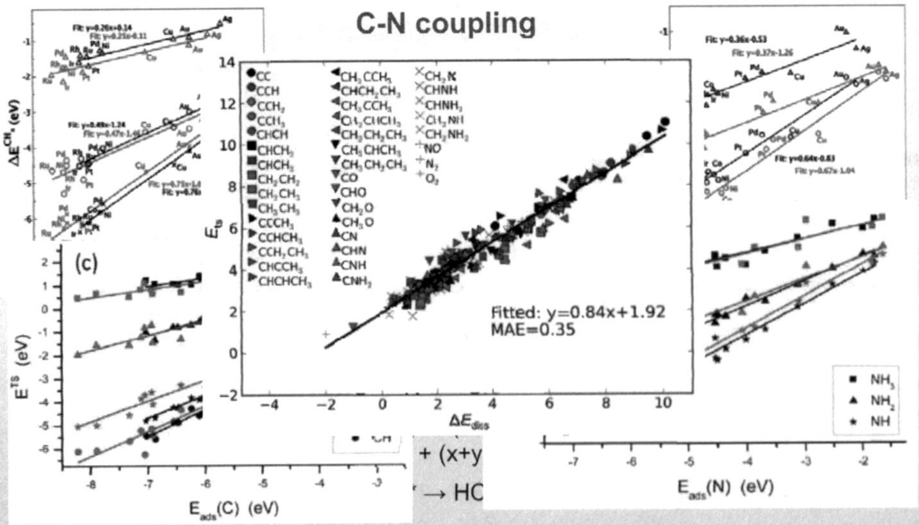

16

AIChE
The Global Home of Chemical Engineers

NH_3 Conversion and Selectivity

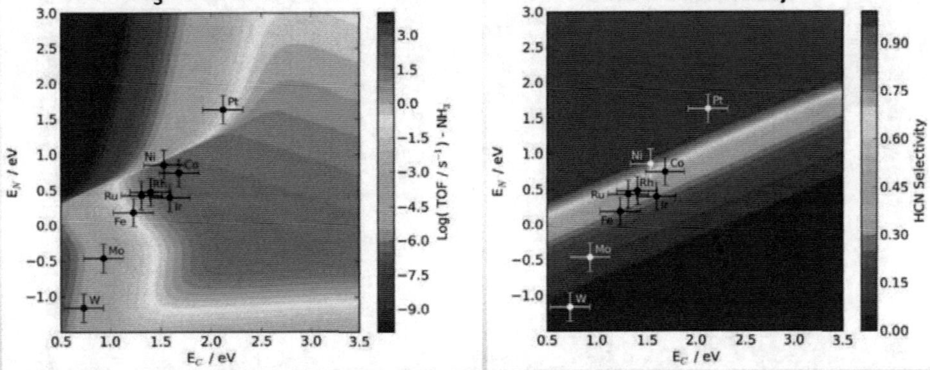

Conditions:
T = 1373 K; 10.7% NH_3; 9.3% CH_4; 15 % conversion wrt CH_4
Descriptor values on step sites of monometallic surfaces are indicated.

Grabow *et al.*, *Angew. Chem. Int. Ed.* **50**, 4601, **2011**

17

AIChE
The Global Home of Chemical Engineers

The Pt Wire Gauze Catalyst

- Pt/Rh gauze catalyst with ca. 10 % Rh (or Ir).
- Rh and Ir increase the long-term durability.
- Activation period of several days.
- Surface restructuring leads to increase surface roughness.
  - Surface bulk diffusion
  - Chemical composition remains constant
- Coke formation can be reduced by adding Al to the catalyst.

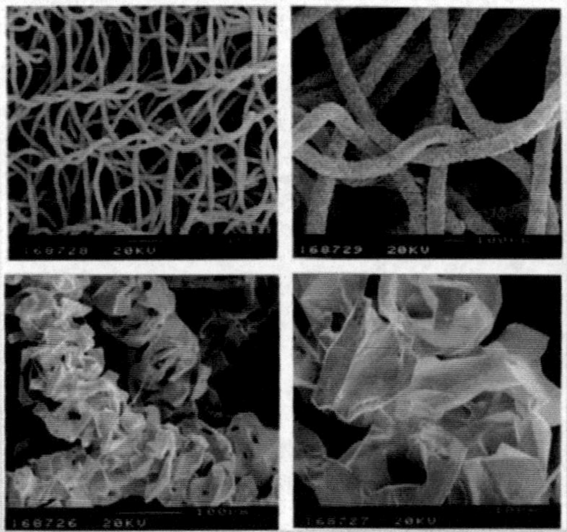

Sauer *et al.* in Handbook of Heterogeneous Catalysis, **2008**

18

C-N Coupling Mechanism

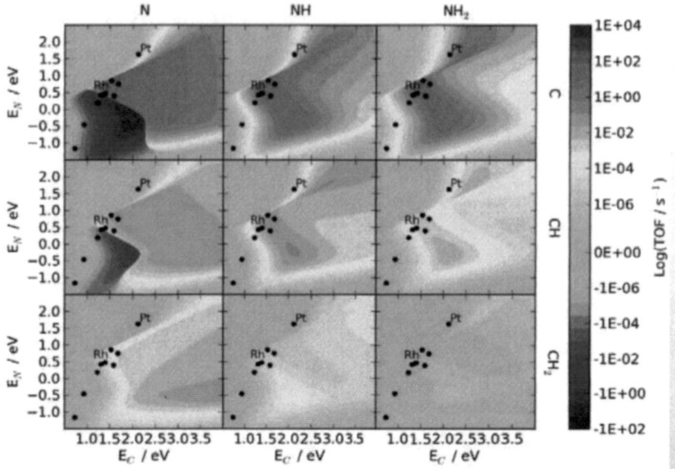

$CH_x + NH_y$ coupling rates at:
1373 K
10.7% NH_3
9.3% CH_4
15 % conversion wrt CH_4

Grabow *et al.*, *Angew. Chem. Int. Ed.* **50**, 4601, **2011**

Comparison to Detailed DFT Study

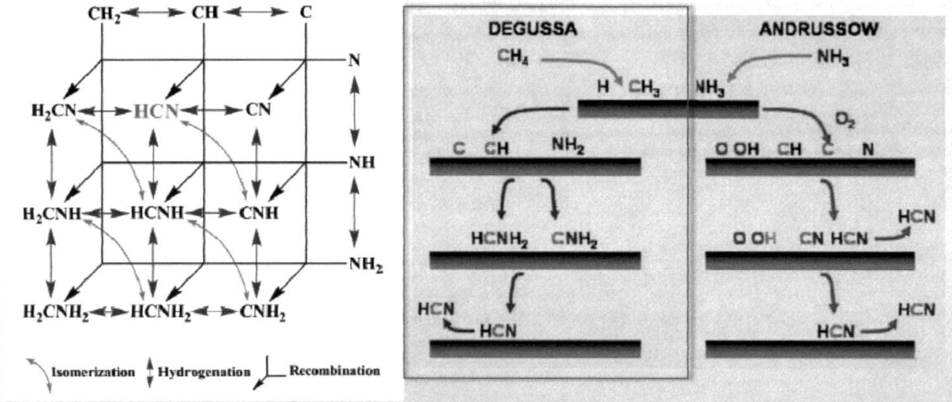

Gomez-Diaz, Lopez, *J. Phys. Chem. C.* **115**, 5667, **2011**

PDS Process Development Symposium

AIChE The Global Home of Chemical Engineers

Recap: Descriptors for Metal Alloys

- Complex catalytic activity can often be described by only a few relevant descriptors.
  - Mechanistic studies may be necessary to identify key intermediates.
  - Existing scaling relations simplify the choice of descriptors.
  - Computational database can be screened for new materials with optimal descriptor values.
- The descriptor-based approach is accurate enough to get a qualitative understanding of complex reaction networks with very little computational effort.
- We addressed the question of the dominant reaction mechanism, activity trends across transition metals and the issue of selectivity in a simple straight-forward approach for an industrial-scale process.

21

PROCESS DEVELOPMENT SYMPOSIUM, JUNE 2-4, 2015, HOUSTON, TX

PDS Process Development Symposium

AIChE The Global Home of Chemical Engineers

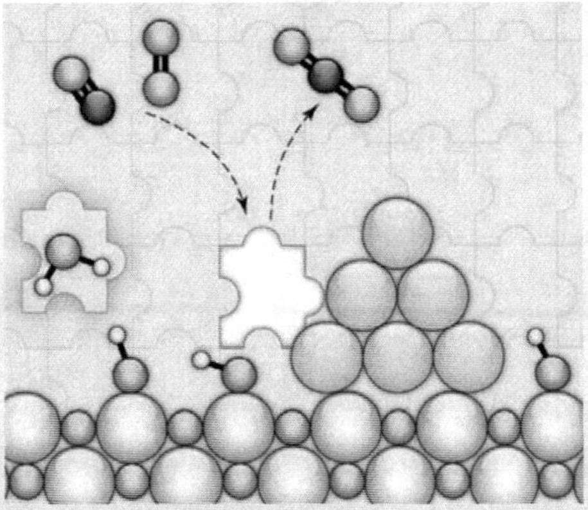

The Puzzle of Gold Catalysis:
CO Oxidation on Au/TiO_2

DESCRIPTION OF METAL/METAL-OXIDE INTERFACES

22

Mullen & Mullins, *Science* **345** (2014) 1564–1565.

PROCESS DEVELOPMENT SYMPOSIUM, JUNE 2-4, 2015, HOUSTON, TX

Water-mediated CO oxidation on Au/TiO_2

- Small amounts of water increase CO oxidation activity on Au
- The reaction occurs at the Au/TiO_2 interface.
- Ti-OH groups anchor water molecules near the interface.
- Water facilitates O_2 activation and extends the reaction zone.
- The rate-determining step involves O-H/D bond breaking.

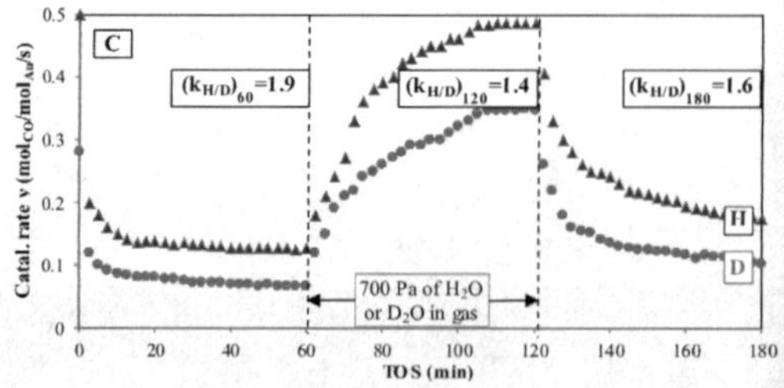

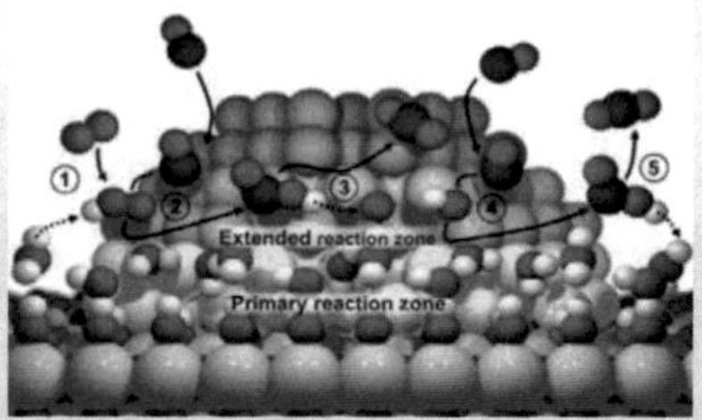

Saavedra, Doan, Pursell, Grabow, Chandler *Science* **345** (2014) 1599-1602

PROCESS DEVELOPMENT SYMPOSIUM, JUNE 2-4, 2015, HOUSTON, TX

23

O_2 Adsorption and Activation

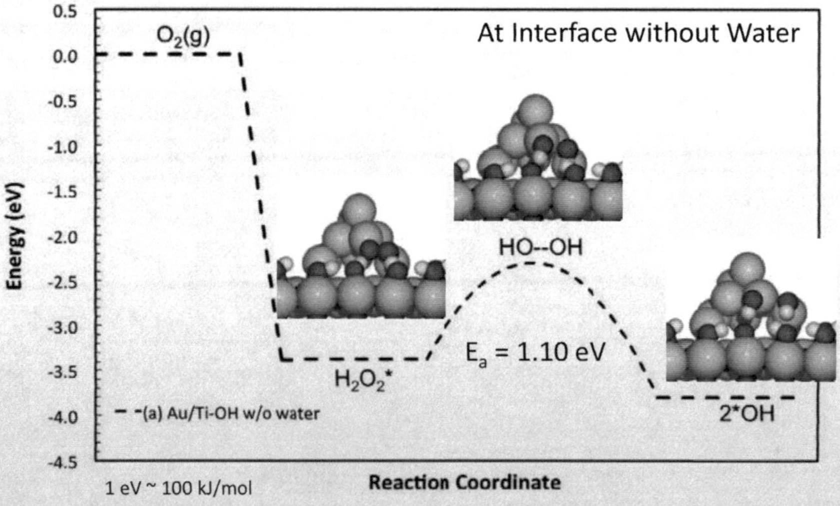

PROCESS DEVELOPMENT SYMPOSIUM, JUNE 2-4, 2015, HOUSTON, TX

24

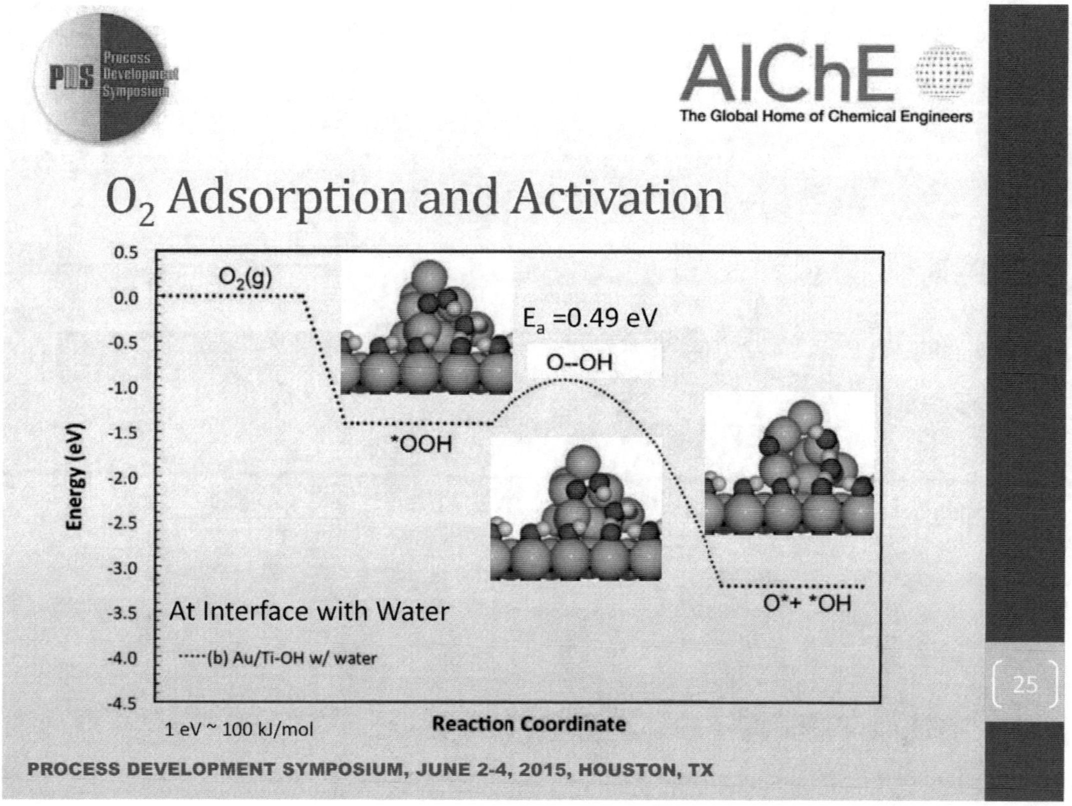
Process Development Symposium
AIChE
The Global Home of Chemical Engineers
O2 Adsorption and Activation
O2(g)
Ea =0.49 eV
O--OH
*OOH
O*+ *OH
At Interface with Water
(b) Au/Ti-OH w/ water
Energy (eV)
Reaction Coordinate
1 eV ~ 100 kJ/mol
PROCESS DEVELOPMENT SYMPOSIUM, JUNE 2-4, 2015, HOUSTON, TX
25

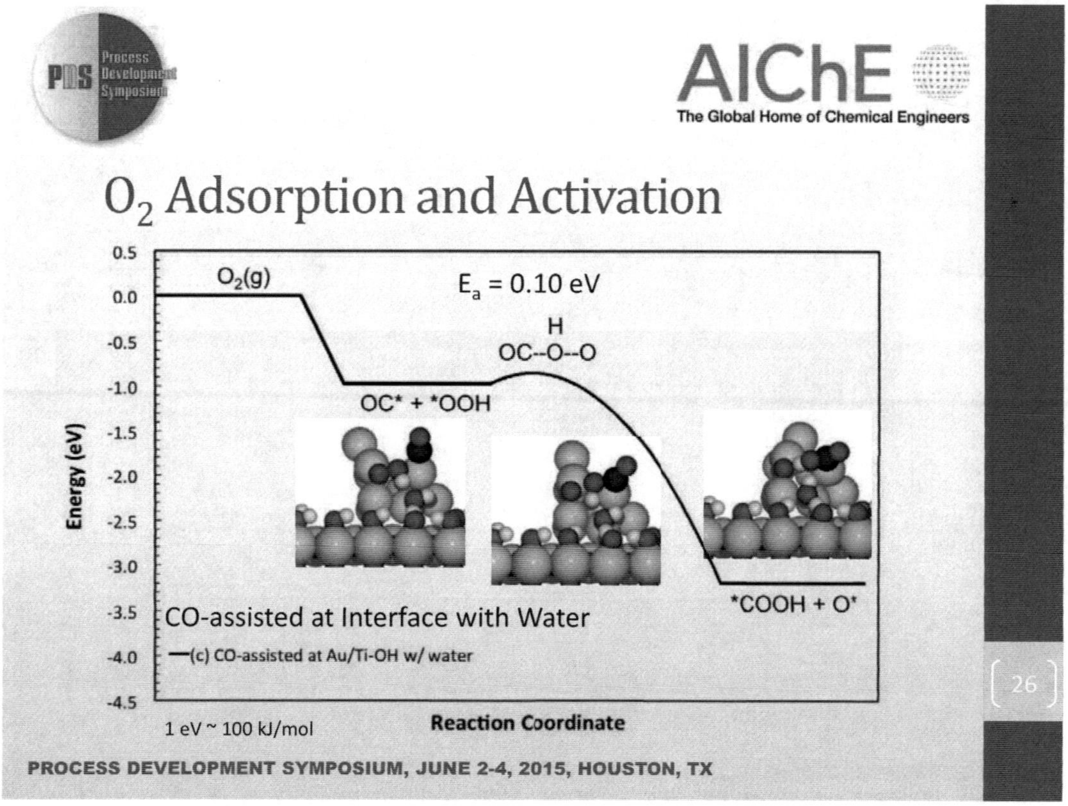
Process Development Symposium
AIChE
The Global Home of Chemical Engineers
O2 Adsorption and Activation
O2(g)
Ea = 0.10 eV
H
OC--O--O
OC* + *OOH
COOH + O
CO-assisted at Interface with Water
(c) CO-assisted at Au/Ti-OH w/ water
Energy (eV)
Reaction Coordinate
1 eV ~ 100 kJ/mol
PROCESS DEVELOPMENT SYMPOSIUM, JUNE 2-4, 2015, HOUSTON, TX
26

The Water-Mediated Mechanism

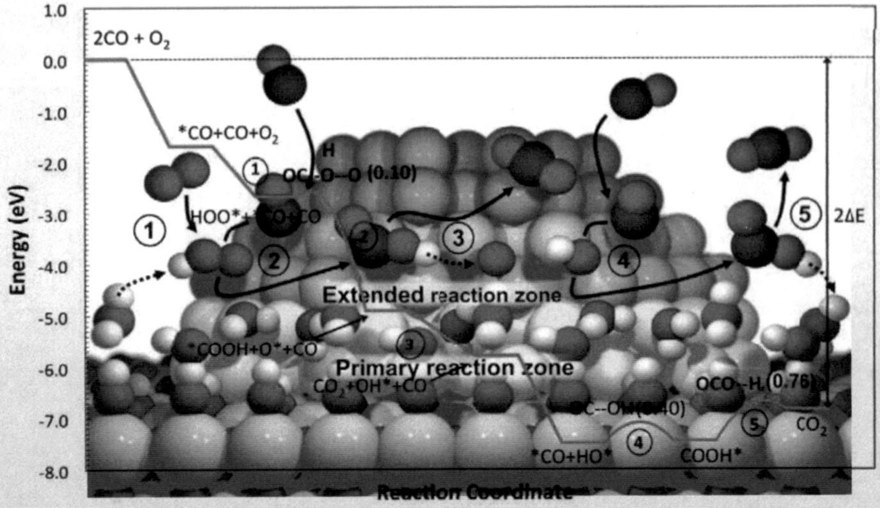

The Rate-Determining Step

PDS Process Development Symposium

AIChE
The Global Home of Chemical Engineers

Summary

- Metal alloy catalysts can be readily screened for activity and selectivity using activity descriptors.
- Metal/metal-oxide interfaces are much more complex:
  - Water acts as a co-catalyst during room temperature CO oxidation at the Au/TiO_2 interface.

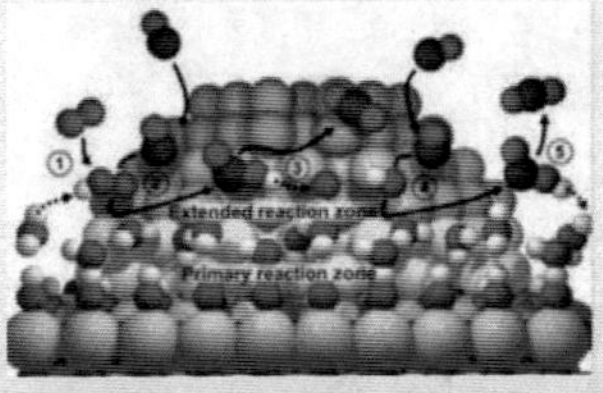

- Benefits of understanding the role of water at the Au/TiO_2 interface:
  - High PROX activity and selectivity by adjusting process parameters (not the catalyst!)
  - Improves selectivity during hydrodeoxygenation (Ru/TiO_2)
  - Potential applications in the mild oxidation of methane (CH_4)

29

PDS Process Development Symposium

AIChE
The Global Home of Chemical Engineers

Acknowledgements

Postdoctoral Researcher: Shengguang Wang

Graduate Students: Byeongjin Baek, Hieu Doan, Arian Ghorbanpour, Sashank Kasiraju, Yuying Song, Melanie Hazlett, Quan Do

Undergraduate Students: Charlie Chirino

Financial support:

- U.S. DOE Early Career Award
- NSF CAREER Award
- NSF/DOE
- Shell, SABIC, Saudi Aramco
- ACS-PRF Young Investigator Award

Computational Resources:

- National Energy Research Scientific Computing Center - NERSC
- Extreme Science and Engineering Discovery Environment - XSEDE (TACC, NICS, SDSC)
- Argonne National Lab (CNM)
- UH Research Computing Center

30

PROCESS DEVELOPMENT SYMPOSIUM, JUNE 2-4, 2015, HOUSTON, TX

PDS Process Development Symposium

AIChE The Global Home of Chemical Engineers

Further Reading

- Computational Catalysis (RSC Catalysis Series)
  Editors: A. Asthagiri and M. J. Janik
  ISBN: 9781849734516
- Tutorial-style with many examples and applications
- Selected Content:
  - Ab-initio thermodynamics
  - First-principles based microkinetic modeling
  - Adaptive kinetic Monte Carlo
  - **Computational catalyst screening**
  - Advances in DFT functionals for catalysis

31

PROCESS DEVELOPMENT SYMPOSIUM, JUNE 2-4, 2015, HOUSTON, TX

HUNTSMAN

Enriching lives through innovation

10 Key Points to Success

JoAnn Lacheny
Huntsman Corporation
Bench Scale Operations Group Leader
AIChE Process Development Symposium
June 2-4, 2015, Houston, TX

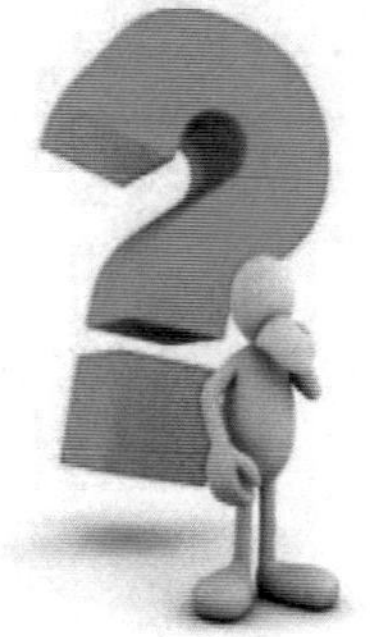

HUNTSMAN

Enriching lives through innovation

- What does success mean to you?

- **Do you need a plan for success?**

HUNTSMAN
Enriching lives through innovation

Definition of Success

Wiktionary.org (24-Feb-2015):

- The achievement of one's aim or goal
- Financial profitability
- One who, or that which, achieves assumed goals

HUNTSMAN
Enriching lives through innovation

✓ **Don't compromise integrity**

➢ Be honest and trustworthy

If it is not Right, Do not do it.
If it is not True, Do not say it.
Marcus Aurelius

HUNTSMAN
Enriching lives through innovation

✓ **Continuous Improvement**
- ➢ Self awareness
- ➢ Learn from your failures

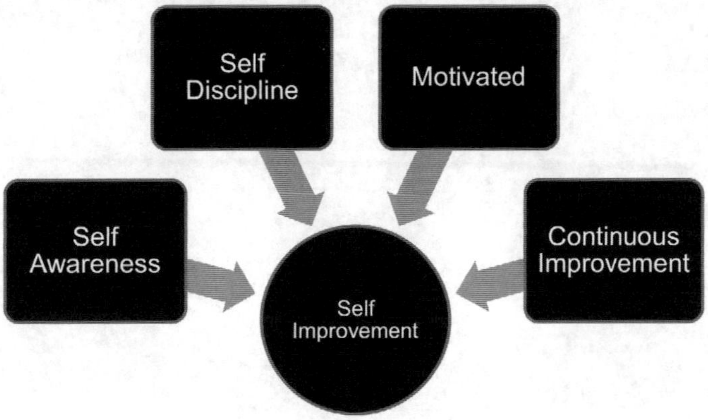

HUNTSMAN
Enriching lives through innovation

✓ **Build relationships**
- ➢ Help others to succeed – you'll build allies.
- ➢ We vs I – share credit
- ➢ Be positive
- ➢ Have compassion
- ➢ Surround yourself with smart people

HUNTSMAN
Enriching lives through innovation

✓ **Be willing to take risks, but also to admit mistakes**

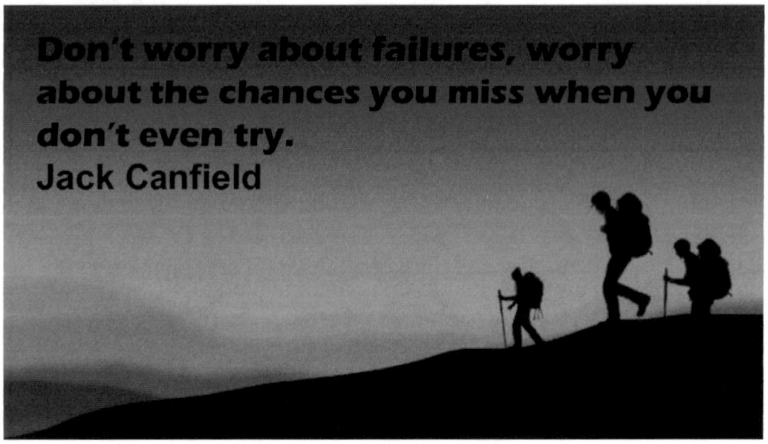

7 *Performance Products*

HUNTSMAN
Enriching lives through innovation

✓ **Work hard and stay focused**

➢ Be thorough, not sloppy

8 *Performance Products*

HUNTSMAN
Enriching lives through innovation

✓ **Be flexible and be willing to wear many hats**
- Be willing to do what others don't want to do
- Broad experience is helpful to seeing the many sides of a problem.
- Change is inevitable, be adaptable

9 *Performance Products*

HUNTSMAN
Enriching lives through innovation

✓ **Communicate…communicate…communicate!**
- Proactively share information; keep others informed
- Listen to others

10 *Performance Products*

HUNTSMAN
Enriching lives through innovation

✓ **Be open to unexpected opportunities**
  ➢ Things you never thought of may come your way.

11 *Performance Products*

HUNTSMAN
Enriching lives through innovation

✓ **Work is serious, but don't take yourself too seriously**

12 *Performance Products*

HUNTSMAN
Enriching lives through innovation

✓ Maintain work-life balance

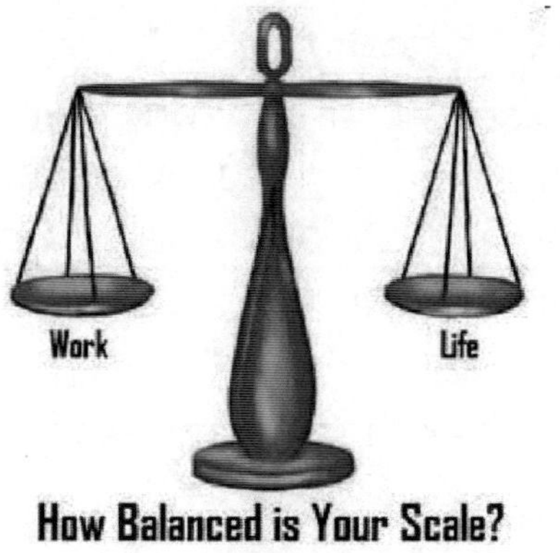

13 Performance Products

HUNTSMAN
Enriching lives through innovation

Take Home Message

- Work hard
- Care about your work
- Value others
- Be adaptable
- Never stop learning

- Realize that your work ethic is a reflection of you as a person.
- Ultimately, success is contributing something good to society in a positive way.
- Greatness is not measured by money, position, or how wonderful you think you are. It is measured by the example you set in how you live your life.

14 Performance Products

- "Coming together is a beginning; keeping together is progress; working together is success." - *Henry Ford*
- "To succeed in life, you need two things: ignorance and confidence." – *Mark Twain*
- "The price of success is hard work, dedication to the job at hand, and determination that whether we win or lose, we have applied the best of ourselves to the task at hand." – *Vince Lombardi*
- "As we look ahead into the next century, leaders will be those who empower others." - *Bill Gates*

THANK YOU

PDS Process Development Symposium

AIChE The Global Home of Chemical Engineers

5 Steps to Your Next Ideal Job

Curtis Swisher

PROCESS DEVELOPMENT SYMPOSIUM, JUNE 2-4, 2015, HOUSTON, TX

PDS Process Development Symposium

AIChE The Global Home of Chemical Engineers

PROCESS DEVELOPMENT SYMPOSIUM, JUNE 2-4, 2015, HOUSTON, TX

PDS Process Development Symposium

AIChE
The Global Home of Chemical Engineers

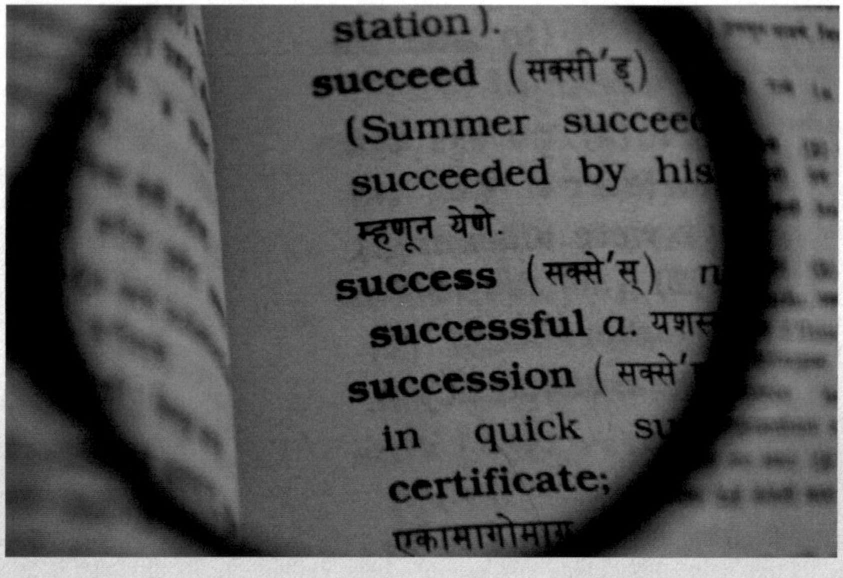

PROCESS DEVELOPMENT SYMPOSIUM, JUNE 2-4, 2015, HOUSTON, TX

PDS Process Development Symposium

AIChE
The Global Home of Chemical Engineers

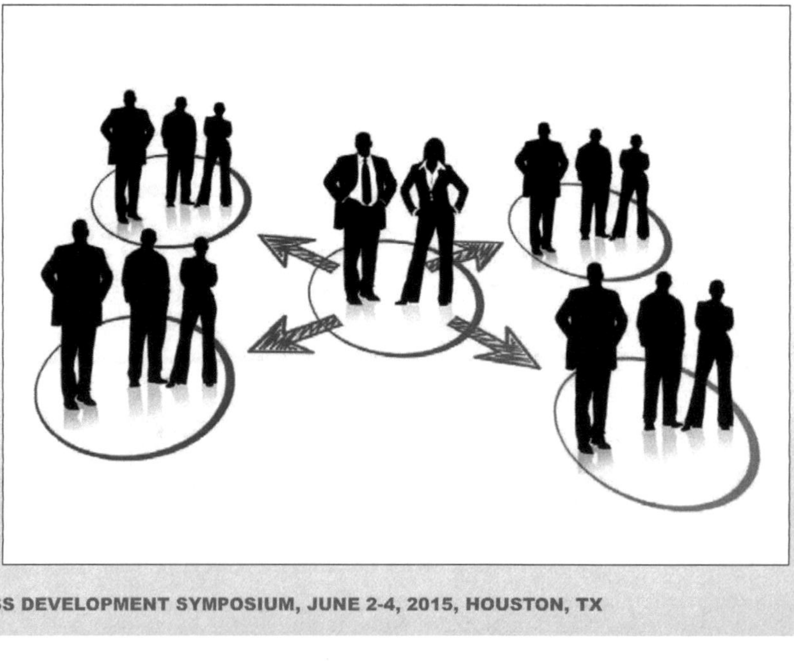

PROCESS DEVELOPMENT SYMPOSIUM, JUNE 2-4, 2015, HOUSTON, TX

PDS Process Development Symposium

AIChE
The Global Home of Chemical Engineers

PROCESS DEVELOPMENT SYMPOSIUM, JUNE 2-4, 2015, HOUSTON, TX

PDS Process Development Symposium

AIChE
The Global Home of Chemical Engineers

PROCESS DEVELOPMENT SYMPOSIUM, JUNE 2-4, 2015, HOUSTON, TX

PDS Process Development Symposium

AIChE
The Global Home of Chemical Engineers

PROCESS DEVELOPMENT SYMPOSIUM, JUNE 2-4, 2015, HOUSTON, TX

PDS Process Development Symposium

AIChE
The Global Home of Chemical Engineers

Early Career Leadership Development

David Van Kleeck
Rice University

PDS Process Development Symposium

AIChE
The Global Home of Chemical Engineers

Who am I?

- BS ChE – Princeton University '73
- PhD ChE – Rice University '81
- 34 Year Career at Shell (R&D)
  - Individual Contributor
  - Manager
  - Campus Recruiter
- 32+ Year Career US Army (Active & Reserve)
  - Commanded units at Battalion, Group, and Brigade
  - Retired as a Brigadier General

PROCESS DEVELOPMENT SYMPOSIUM, JUNE 2-4, 2015, HOUSTON, TX

Resisting Learning Leadership Skills is Futile

You must understand the fundamentals of leadership even if you never intend to become a manager

PROCESS DEVELOPMENT SYMPOSIUM, JUNE 2-4, 2015, HOUSTON, TX

Strategic Level Challenges

- Pace of Change
- Ethical Dilemmas.

Leadership Imperative: Adapt and learn

PROCESS DEVELOPMENT SYMPOSIUM, JUNE 2-4, 2015, HOUSTON, TX

PDS Process Development Symposium

AIChE
The Global Home of Chemical Engineers

Operational Level Challenges

- Globalization
- Information Management.

Leadership Imperative: Organize work processes

PROCESS DEVELOPMENT SYMPOSIUM, JUNE 2-4, 2015, HOUSTON, TX

PDS Process Development Symposium

AIChE
The Global Home of Chemical Engineers

Tactical Level Challenges

- Interpersonal Skills
- Goal Setting
- Project Management.

Leadership Imperative: Learn to lead yourself – be proactive

PROCESS DEVELOPMENT SYMPOSIUM, JUNE 2-4, 2015, HOUSTON, TX

Planning Leadership Growth

- Self Assessment
  - Personality
  - Core Values
  - Stress Management
- Motivation to Lead
- Manage Competency Gaps

PROCESS DEVELOPMENT SYMPOSIUM, JUNE 2-4, 2015, HOUSTON, TX

Spectrum of Leadership Skills

- Covey Framework
  - Dependence
  - Independence
  - Interdependence
- Leadership Competencies – Company Specific
- Competency Skill Level versus Experience

PROCESS DEVELOPMENT SYMPOSIUM, JUNE 2-4, 2015, HOUSTON, TX

PDS Process Development Symposium

AIChE The Global Home of Chemical Engineers

RCEL Competency Framework

- Personal Domain
  - Setting and Achieving Goals
  - Managing Stress
  - Problem Solving and Decision Making
- Interpersonal Domain
  - Managing Conflict
  - Building Positive Relationships
  - Oral and written Communication
- Management Domain
  - Organizing/Staffing
  - Providing Feedback
- Leadership Domain
  - Motivating and Inspiring Others
  - Leading Change

Dependence ↓ Independence ↓ Interdepence

PROCESS DEVELOPMENT SYMPOSIUM, JUNE 2-4, 2015, HOUSTON, TX

PDS Process Development Symposium

AIChE The Global Home of Chemical Engineers

Bloom's Taxonomy of Skills

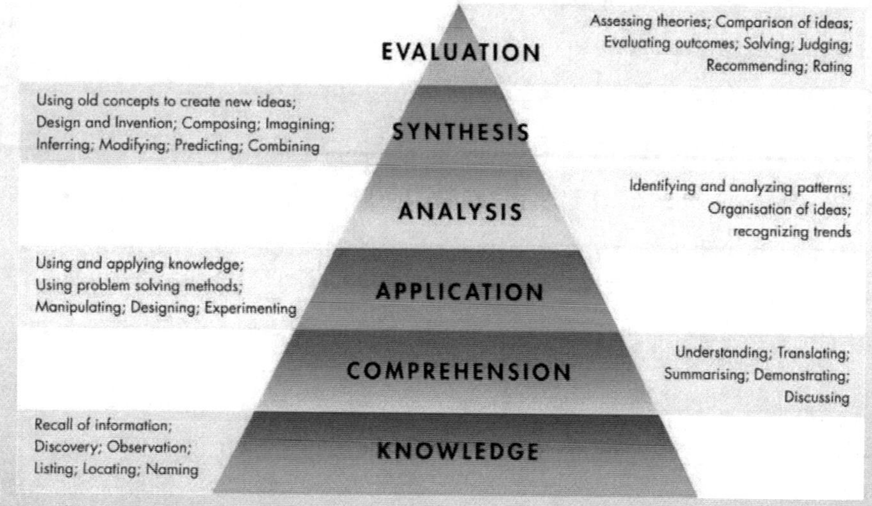

PROCESS DEVELOPMENT SYMPOSIUM, JUNE 2-4, 2015, HOUSTON, TX

Competency Map/Gap Identification

- Gaps are the difference between current assessment and expected competency level for specified job skills
- Assessments should be the result of a 360 degree feedback analysis in addition to self assessment
- Specific measures for establishing competency are extremely important to the analysis
- Actions to close the gaps are incorporated in a SMART goals format to drive action

Leadership Development Plan

- Should be a formal document
- Captures as Goals
  - Additional Training Needed
  - Job Assignments Needed
- Lists Resources Needed to complete the plan
- Should be viewed as a "Contract" between the boss and the employee
- Should have both a short term and long term focus
  - Sort Term = 1 year
  - Long Term = 3-5 years

PROCESS DEVELOPMENT SYMPOSIUM, JUNE 2-4, 2015, HOUSTON, TX

PDS Process Development Symposium

AIChE The Global Home of Chemical Engineers

Leadership Growth Pattern

- Dependent stage of development
  - Can manage own work
  - Can set own goals with coaching
- Independent stage of development
  - Can manage a small work group
  - Can set team goals
  - Can provide feedback to others
- Interdependent stage of development
  - Can manage large teams or multiple teams
  - Can set a strategic direction for team efforts
  - Can coach others to perform at a high level

PROCESS DEVELOPMENT SYMPOSIUM, JUNE 2-4, 2015, HOUSTON, TX

PDS Process Development Symposium

AIChE The Global Home of Chemical Engineers

Career Planning

- Role of the direct manager/supervisor
  - Coach employee performance
  - Support the Leadership Development Plan
  - Structure job assignments to support the Development Plan
- Role of the manager's manager
  - Mentor the employee
  - Provide career development guidance
  - Assess potential for advancement

PROCESS DEVELOPMENT SYMPOSIUM, JUNE 2-4, 2015, HOUSTON, TX

PDS Process Development Symposium

AIChE
The Global Home of Chemical Engineers

Leadership is not a spectator sport

- The only way to master leadership skills is to perform them even if only poorly at first
- Taking on leadership roles is inherently risky. To be successful, the leader must accept some risk.
- The effective leader has a purpose and direction. It does not typically come from the followers.

PROCESS DEVELOPMENT SYMPOSIUM, JUNE 2-4, 2015, HOUSTON, TX

Mid-Atlantic Technology, Research & Innovation Center

MATRIC

Market-Driven Innovation

MATRIC | David Statler, Ph.D. | June 4th, 2015

Concept to Commercialization: Methodology to Sustainable Process Success

CHEMICAL, ENERGY & ENVIRONMENTAL TECHNOLOGIES

TECHNICAL ENGINEERING

ADVANCED SOFTWARE TECHNOLOGIES

Agenda

- **About MATRIC**
- **Laboratory to Commercialization Success**
- **Sustainable Process Design**
  - Objectives
  - Challenges
  - Work Process Methodology for *Sustainable Design*
  - Concept to Commercialization

Sustainable Design: applicable to all designs.

- Renewables or Non Renewables
- Existing or New Processes

MATRIC

2

Mid-Atlantic Technology Research & Innovation Center

- Strategic innovation partner that provides deep, uncommon expertise and infrastructure to solve the most challenging problems in the areas of science and technology.

- With expertise in chemical, energy and environmental technology, and advanced software, MATRIC enables its partners to move ideas from conception to market with speed and efficiency.

MATRIC

3

From the Lab to Commercialization

2009 Laboratory

2010-13 Pilot Plant

2014 Commercialized!

MATRIC

4

Commercialization, Concept

- Stakeholders
- Sponsors Expectations
- End Goal In Mind
- Technology Requirements
- Inherent Safety

- What is Success?

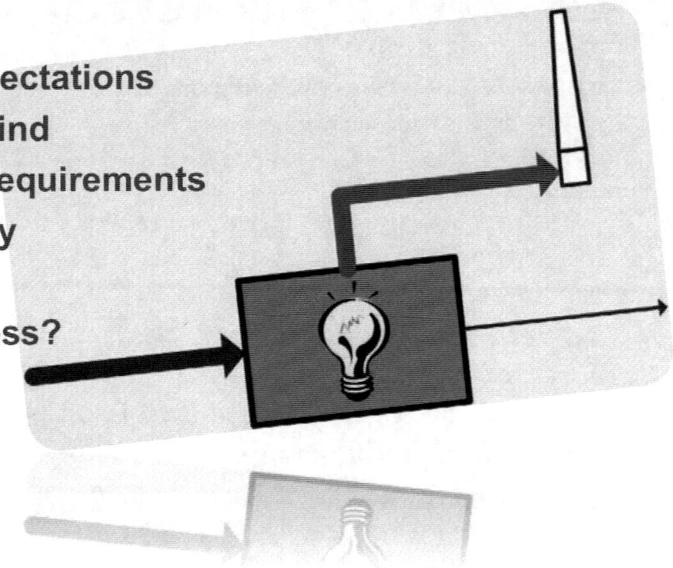

Mid-Atlantic Technology, Research & Innovation Center
MATRIC

5

Sustainable Process

- Sorbent
  - Green
  - Non-hazardous
  - Long Life, years…
  - Recycled over and over
    - Much better than current consumable technology
- Energy Efficiency
  - Energy Recovery
  - Waste Minimized
  - Low Energy Use through tight parameter control

Mid-Atlantic Technology, Research & Innovation Center
MATRIC

6

Sustainable Process Design Objectives

- **What are your process design objectives and where should you look for sustainable design improvements or optimization?**
  - New / alternative raw materials *(toxic to non-toxic, renewable, green)*
  - Alternative process chemistry *(replace intermediates with non-toxic ones)*
  - Alternative and flexible process design *(Pretreatment, RO in place of distillation)*
  - Recoverable waste streams *(less water in waste, less toxics)*
  - Energy optimization *(Integration, Pinch Point)*
  - New end product *(Changing Specifications)*
- **Drivers (existing situation, customer, market, political changes)**
- **Continuous Improvement and Process Development**
- **Economics always wins**

Mid-Atlantic Technology, Research & Innovation Center
MATRIC

7

Sustainable Process Design Challenges

- **Commercialization (get to commercial scale, economically...)**
- **Feedstock**
  - Impurities
  - Seasonality
  - Market cycles
  - Location, sources
- **Process Steps**
  - Product yield
  - Dilute Products: separations, energy, complexity.
  - Hazardous materials: substitution, elimination, and reduction.
  - Product quality- impurities and byproducts
- **The final process design must be equal or more economical than the competition.**

Mid-Atlantic Technology, Research & Innovation Center
MATRIC

8

Sustainable PD (A Work Process)

- **Regardless of where you start, sustainable process designs require a diverse thought process and a broad understanding of the value chain.**
- **Experienced team with diverse backgrounds are key.**
  - Chemistry alternative knowledge
  - Process alternatives knowledge
  - Market knowledge
  - Techno-Economic knowledge

- **Infrastructure is necessary**
  - Validating the chemistry and verifying process feasibility.
- **Systematic, and step wise approach with simulation and economics**
- **Discipline methodology of risk reduction**

Mid-Atlantic Technology, Research & Innovation Center
MATRIC

9

Commercialization, Concept to Lab

- **In the Lab**
  - Understand and Validate the Chemistry
    - Test concepts and verify in lab
    - Do not assume the process chemistry free of byproducts
    - Understand the chemistry and the byproducts
  - Material & Energy Balances:
    - Concentrations (products)
    - Separations (byproducts)
    - Waste streams
    - Energy demand
  - Feed the Model
  - Check the Economics

Mid-Atlantic Technology, Research & Innovation Center
MATRIC

10

Commercialization, Lab to Pilot Plant

- **Develop the process**
  - Piloting = proof testing and verifying feasibility of concept.
  - Validate the model.
  - Begin piloting of the major process steps with a *prioritization according to risk.*
  - Skip the known technologies until demonstration is required; focusing on most risky first.
  - Close the recycle loops. Assumptions can be way off.
- **Back to the Model**
  - Check and update the material and energy balance
- **Back to the Economics.**
  - Is the process advantaged?
  - Back to the drawing board?
  - On to process optimization?

Mid-Atlantic Technology, Research & Innovation Center
MATRIC

11

Process Hurdles

- **Recycle loops demonstrated in the Pilot Plant is essential**
  - Study side chemistry and non-reversible by-products
  - Study rates of accumulation
  - How to handle by-product
- **Energy Efficiency is challenge to demonstrate at PP level**
  - Heat Interchangers incorporated
  - Lab Scale and PP have greater heat losses due simply to size
  - Simulations a must

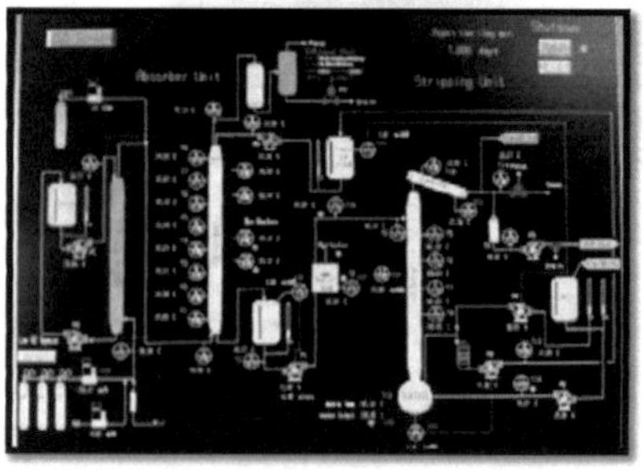

Mid-Atlantic Technology, Research & Innovation Center
MATRIC

12

Commercialization, Process Optimization & Continuous Process Improvement

- By-product reactions rates
- Slip Streams
- Energy minimization
- Parameter control
- Corrosion rates
- Pre-treatment
- Optimization

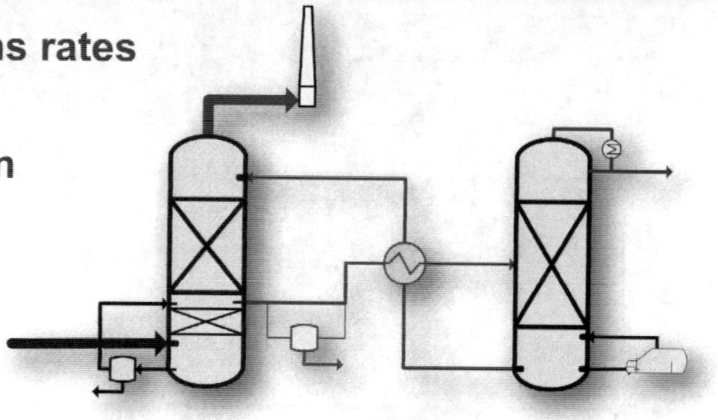

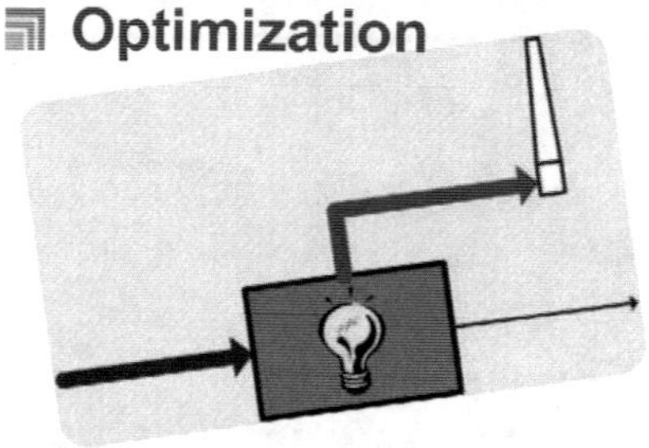

MATRIC

13

Sustainable Process Design

- Commercialization
  Demands it.
- Economics
  Commands it.
- Continuous Improvement
  Lands it.

MATRIC

14

Sustainable Process Success

Special Thanks to Mark Nunley! of MATRIC for his help with this presentation.

Thanks for your kind attention

MATRIC

15

AIChE
The Global Home of Chemical Engineers

Challenges for the Chemical Industry

John G. Pendergast
Senior Fellow: Oil and Gas, Advanced Materials
Formally: Fellow Process Separations Engineering & Process Sciences
The Dow Chemical Company

PROCESS DEVELOPMENT SYMPOSIUM, JUNE 2-4, 2015, HOUSTON, TX

AIChE
The Global Home of Chemical Engineers

Outline

- Dow Introduction
  - Company profile
  - Contributions and drivers
- Historical Perspective
  - Where do we come from
- Our Industry in Perspective
- Challenges
  - With at least one suggestion for direction
    - Bio-processing, chemicals from bio-processing √
    - Distributed (small scale) manufacturing √
    - Retrofit of existing processes√
    - CO_2 Capture and Sequestration √
    - Methane activation √
    - Olefin recovery and separation √
    - Low level heat recovery √
    - New Realities √
    - Demographics and expertise development
- Questions, discussions

PROCESS DEVELOPMENT SYMPOSIUM, JUNE 2-4, 2015, HOUSTON, TX

The Dow Chemical Company

AIChE
The Global Home of C

- Diversified chemical company, harnessing the power of science and technology to improve living daily

1897	2000	2008	Olin
Founded Midland, MI	Acquired Union Carbide	Acquired Rohm & Haas	

Dow by the numbers

- 35 Countries
- 188 Sites
- 5000+ Products
- 50,000 WW Employees
- 60++ Billion Annual Sales

Unique perspective of commodity landscape and specialty chemicals

Founded on brine chemistry 117 years ago

Large Scale
Ethylene and polyethylene
Chlorine and caustic
Ethylene oxide and ethylene glycol

Specialty Scale
Metal organics, tri-methyl gallium, tri-methyl indium
OLED materials
Chemical mechanical planarization products, photoresists...

PROCESS DEVELOPMENT SYMPOSIUM, JUNE 2-4, 2015, HOUSTON, TX

Large Integrated Manufacturing Sites

AIChE
The Global Home of Chemical Engineers

Plaquemine, U.S.A.

Terneuzen, Netherlands

Stade, Germany

Freeport, U.S.A.

PROCESS DEVELOPMENT SYMPOSIUM, JUNE 2-4, 2015, HOUSTON, TX

Sustainability is Crucial to Our Success

AIChE
The Global Home of Chemical Engineers

3,700 MW of electricity required to operate Dow plants	=	3,700 MW of electricity Used by San Francisco, San Diego, & Oakland combined

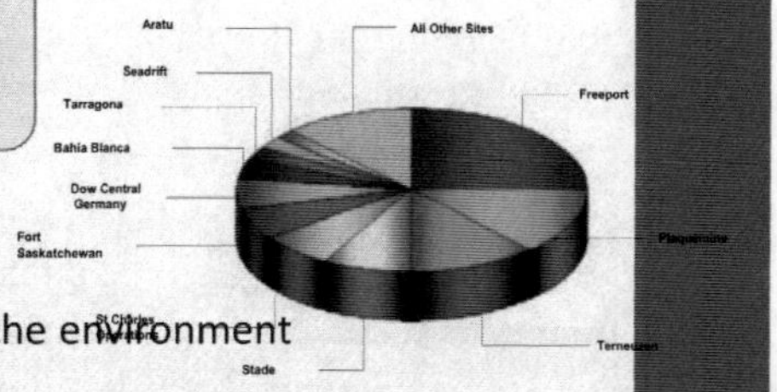

- Feedstock demand is **800,000 barrels/day**
- Leading innovator in cogeneration
  - Increased efficiency with reduced impact on the environment
  - Uses **20-40% less fuel**
- Self-generates **~75%** of all power & steam
- Operates over **$6.2 billion** in energy assets & supports **$2.5 billion** in JV assets
- Operate our own Wastewater treatment
- Operate our own incineration facilities
- We strive to be good stewards because it is good business

PROCESS DEVELOPMENT SYMPOSIUM, JUNE 2-4, 2015, HOUSTON, TX

ASEE: For External Release

Sustainable Chemistry

AIChE
The Global Home of Chemical Engineers

- Integrated look at chemicals, processes and products
  - Improve safety of manufacturing processes
  - Maximize energy efficiency
  - Maximize the efficiency of raw material use
  - Minimize the use of scarce resources
  - Minimize the return of anthropogenic substances to nature
  - Minimize environmental impact
  - Significantly enhance benefits to society
- Will change the way that some products are made
  - Sustainability is good business
- It's all about balance
- **Our products must be profitable**
  - **Alternatives must be supported financially**
  - **Alternatives compete for precious capital**

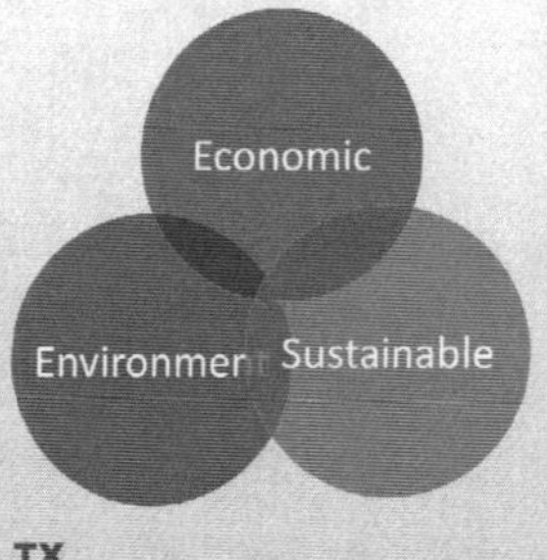

PROCESS DEVELOPMENT SYMPOSIUM, JUNE 2-4, 2015, HOUSTON, TX

ASEE: For External Release

Bio-based Products

AIChE
The Global Home of Chemical Engineers

- Bio-based product are challenged to compete on a cost basis with hydrocarbon based products
  - The consumer has a very limited willingness to pay a premium for "green products"
    - In products far removed from the end user (commodity) , the consumer has even less willingness to pay
  - Several notable examples of this difficulty
    - Poly-lactic acid
    - Chitin and chitin derived materials
  - The lesson: simply providing a "renewable product" does not guarantee acceptance
    - Certainly does not guarantee financial success
- Specialty chemicals, in our opinion, have promise for bio-derived materials
  - Higher margins, less commodity mentality
  - Some tolerance for a premium

PROCESS DEVELOPMENT SYMPOSIUM, JUNE 2-4, 2015, HOUSTON, TX

Distributed Manufacturing

AIChE
The Global Home of Chemical Engineers

- Challenge
  - We must be realistic ... distributed manufacturing is a huge challenge
    - Seeks to reverse ~100 years of consolidation
  - Design philosophy
    - Simply making a plant smaller will not work
    - Design philosophy must be different
    - This is in sharp contrast to the paradigm shown earlier
  - Safety must be a major consideration / concern
- Distributed manufacturing may be a partial solution to effective use of biomass
  - Resource radius limitations
    - Biomass cannot economically be shipped past a certain radius
    - 80 – 100 kilometers
  - Stranded resources
    - No pipeline, gas flaring (small scale GTL)
  - Distributed manufacturing may be used to reduce some transportation hazards
    - Drive to reduce "rail miles" for many chemicals
    - Some materials are too hazardous to routinely ship
    - Still must take place is a professional, tightly monitored environment
  - Political instability
    - Move your asset (Plant on a ship)

PROCESS DEVELOPMENT SYMPOSIUM, JUNE 2-4, 2015, HOUSTON, TX

Maintaining Efficiency Across Lifecycle

AIChE The Global Home of Chemical Engineers

- Our processes (physical facilities) have <u>long</u> life spans
  - 30 years is not uncommon ... 50 years not unknown
    - Maintenance is performed and parts replaced
    - Instrumentation upgraded
    - Plants are by no means neglected .... <u>but</u>
  - Basic configuration may not have changed ... in two generations!
- How do we find and implement process alternatives to <u>improved</u> that meet our economic criterion
  - Energy cost alone <u>will not</u> support total replacement
  - Additional capacity generally will support capital expenditure
  - It must fit into the existing framework
  - We must uncover "all" alternatives
  - Dow has done a great deal of work in Synthesis and Intensification
    - Professor Rakesh Agrawal Professor Grossman have done some very interesting work in the area of separations, as have others

PROCESS DEVELOPMENT SYMPOSIUM, JUNE 2-4, 2015, HOUSTON, TX

Separation Sequencing

AIChE The Global Home of Chemical Engineers

- We have the tools and the techniques to
    - Generate <u>all</u> sequences
    - Simple <u>plus</u> complex
    - Heat integrated sequences
  - Use optimization tools to decide the "best" sequence
- Our challenge
  - Use these tools to build new plants
  - Even more challenging: Use these tools to find better sequences that fit our <u>existing</u> facilities

Extremely challenging but very important

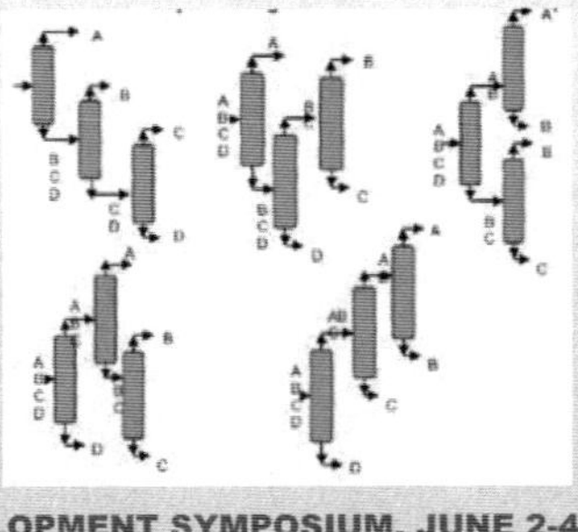

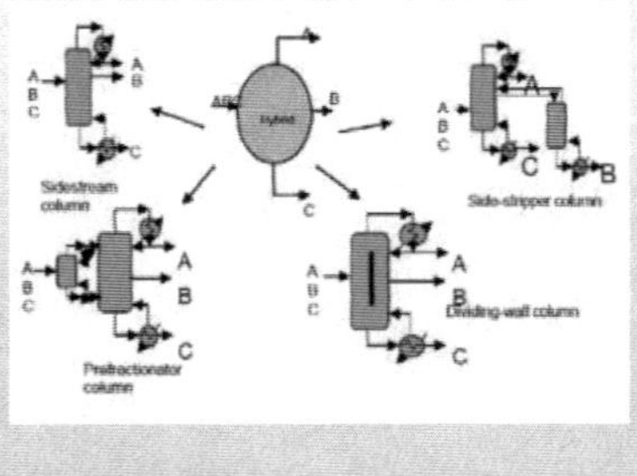

PROCESS DEVELOPMENT SYMPOSIUM, JUNE 2-4, 2015, HOUSTON, TX

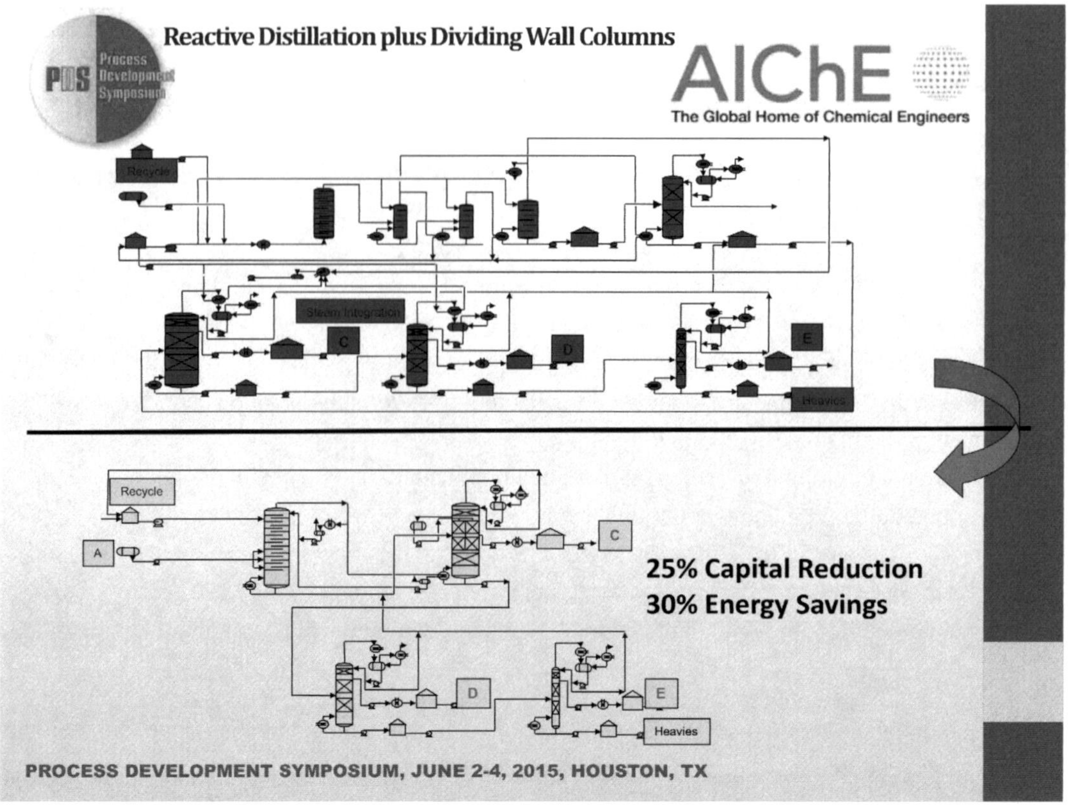
Reactive Distillation plus Dividing Wall Columns
PDS Process Development Symposium
AIChE
The Global Home of Chemical Engineers
Recycle
Steam Integration
C
D
E
Heavies
Recycle
A
C
D
E
Heavies
25% Capital Reduction
30% Energy Savings
PROCESS DEVELOPMENT SYMPOSIUM, JUNE 2-4, 2015, HOUSTON, TX

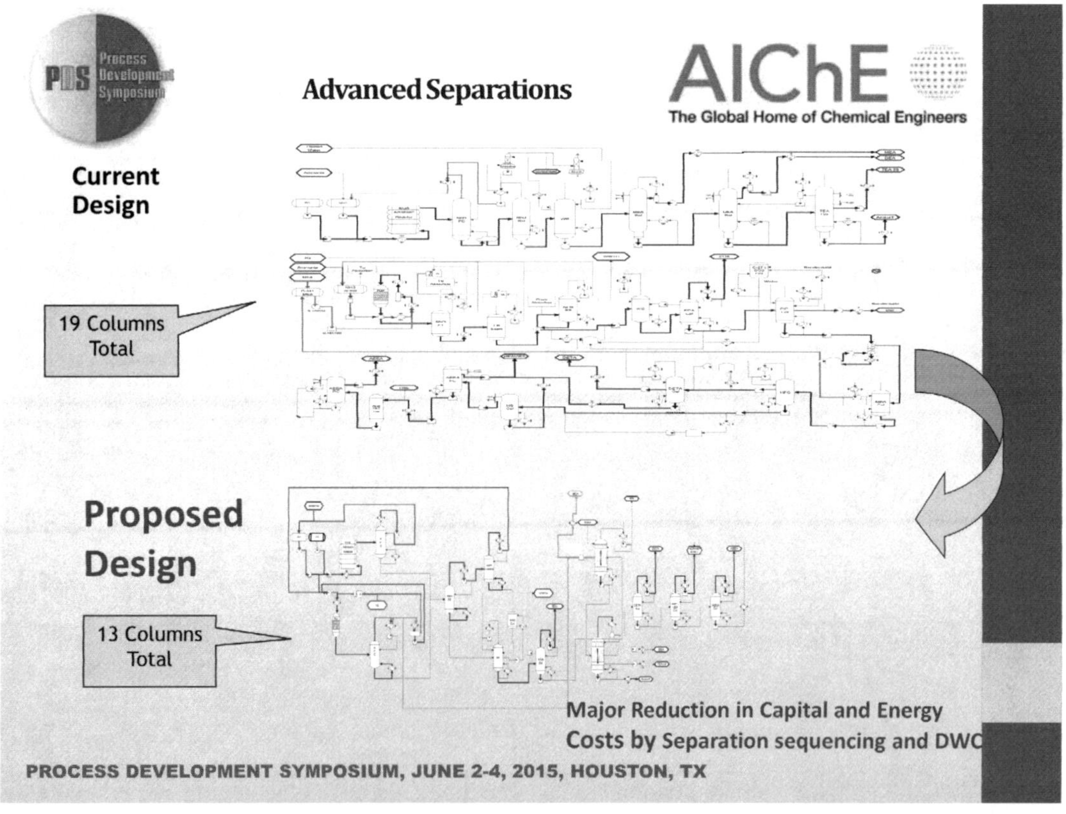
PDS Process Development Symposium
Advanced Separations
AIChE
The Global Home of Chemical Engineers
Current Design
19 Columns Total
Proposed Design
13 Columns Total
Major Reduction in Capital and Energy
Costs by Separation sequencing and DWC
PROCESS DEVELOPMENT SYMPOSIUM, JUNE 2-4, 2015, HOUSTON, TX

Reactive Distillation

AIChE The Global Home of Chemical Engineers

Where applicable, one of the most powerful intensification techniques available

- Reaction must take place at approximately the same conditions as the separation
- Generally thought of as being applicable to reacting away azeotrope
- Generally thought of as driving equilibrium limited reactions

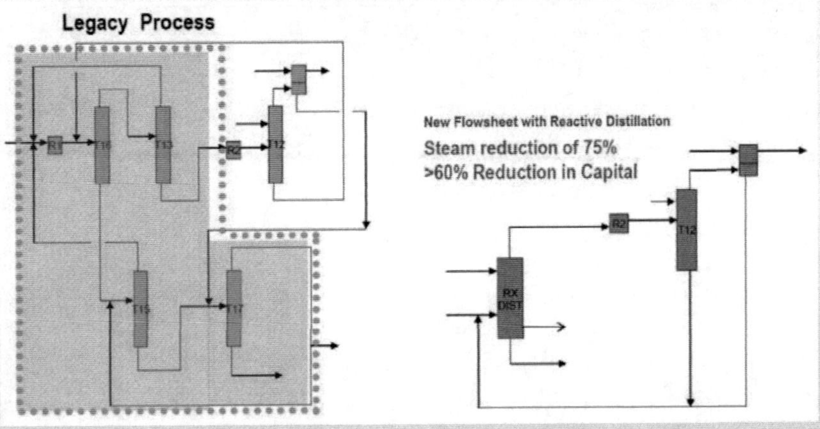

PROCESS DEVELOPMENT SYMPOSIUM, JUNE 2-4, 2015, HOUSTON, TX

CO2 Capture and Sequestration (CCS)

AIChE The Global Home of Chemical Engineers

The Chemical Process Industry <u>is not</u> the bulk of the problem

- Our industry is not trivial
- Depending on the accounting method 5% (1) of global CO_2
  - Many of our larger products produce <u>roughly</u> one mass unit of CO_2 per mass unit of product
- We are in an excellent position (perhaps the best) to do our share

- CCS <u>is not</u> a technical issue
  - CCS is an <u>economic</u> issue
    - Credible estimates \$70 - \$200 per ton
    - We compete in a global market
    - Unilateral application puts us at an economic disadvantage
  - CCS is a litigation / risk management issue

- Potential fruitful areas of development
  - Enriched oxygen (membranes / PSA) and CO_2 capture on furnace vents
    - Reduces flow requirements
  - Optimize fractional recovery for economic benefit
    - Is 90% capture the optimum number ?

(1) Technology Roadmap Energy and GHG Reductions in the Chemical Industry via Catalytic Processes 2013 Rights IEA

PROCESS DEVELOPMENT SYMPOSIUM, JUNE 2-4, 2015, HOUSTON, TX

Methane Activation

AIChE
The Global Home of Chemical Engineers

- Direct coupling of methane without syn-gas
  - In the interest of full disclosure: Many companies believe in methane to products through syn-gas
    - Fischer–Tropsch process
    - Methanol to Gasoline process (MTG)
    - Syngas to gasoline plus process (STG+)
    - Dow process US8129436B2
      - Our focus is different because our focus is C_2=, C_3=
- Direct and / or Oxidative Coupling of Methane (OCM)
  - The Holy Grail of Hydrocarbons
  - Fundamental problem: the intermediate is more reactive than either the reactant or final product
  - Conventional wisdom says 30-40% conversion to C_2 is the minimum for economic viability

Methane Activation

AIChE
The Global Home of Chemical Engineers

- How do you change the paradigm?
  - Fractionation under current conversions is not practical
- Remove the intermediate olefin and concentrate the olefin for purification
  - Membrane
    - Larger species permeating
    - High flux more critical than high selectivity
  - Reactive absorption
    - Selectivity, safety and size are key issues

Some recent and encouraging work on methane activation.
Too early to tell if real or hype

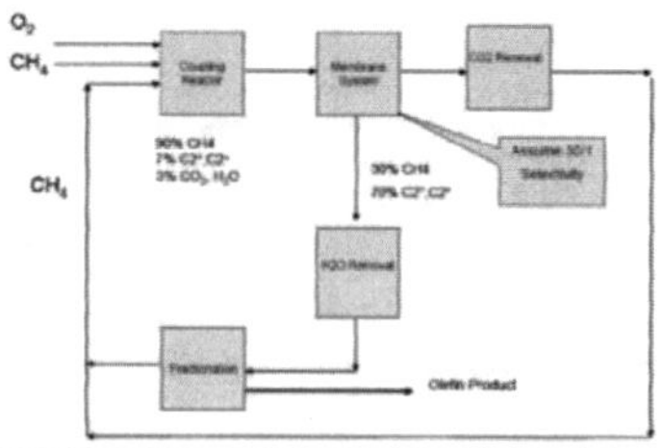

Olefin Recovery and Separation

AIChE
The Global Home of Chemical Engineers

- Olefin manufacturing dominated by separation landscape
  - Risk aversion is enormous ... and understandable
  - Cryogenic separation rule
  - Separation of olefin product away from cracked gas would change paradigm
- Can also be extended to olefin recovery from process vents
  - Recycle
    - Places false load on ethylene plant
    - High risk of contamination and / or fouling
  - Adsorption, absorption, membranes
    - Often economically challenged
- Potentially fruitful area
  - Reactive absorption
    - There are safety considerations that must be considered
      - Scales well
        - Could open the door for smaller, more compact olefins production
        - This will be an uphill struggle
    - Facilitated membrane transport

Low Level Heat Recovery

AIChE
The Global Home of Chemical Engineers

- Large facilities generally have large quantities of low level heat
  - Process facilities are generally well integrated with respect to high level heat
    - Within unit boundaries ... reactors to separations
    - Across facilities ... electricity, boilers, steam generation, boiler feed water ... etc.
  - Condensing temperature limitations
  - Equipment limitations
- Many techniques known for low level heat recovery...
  - Difficult to deliver economic justification
    - In a strange twist, improved energy cost make justification more difficult!
    - May need justification from other sources ... carbon credits ... etc.
- Well known methods (1)
    - Direct utilization ... drying, heating
    - Vapor Compression Heat Pump
    - Organic Rankine
    - Thermo-compressor
    - Refrigeration Cycles
- Utilization of absorption refrigeration cycles
  - Integration with site infrastructure
    - Lower pressure in equipment ... improved safety
    - Use on satellite areas
  - Provide to commercial / residential as chilled water
    - Fraught with litigation concerns ... cross-contamination risks may make this impossible
- Overarching concern
  - What is our failsafe position?
  - If we design our cooling water for 7-10 C, what happens if the chillers fail?

(1) Ankur Kapil, Igor Bulatov, Robin Smith, Jin-Kuk Kim *Advanced Process Integration for Low Grade Heat Recovery*

PROCESS DEVELOPMENT SYMPOSIUM, JUNE 2-4, 2015, HOUSTON, TX

PDS Process Development Symposium

New Realities

AIChE
The Global Home of Chemical Engineers

- Water consumption
  - Historically, simulations have not even counted water
  - Large integrated sites consumption
  - Are we taking water from farms, urban areas?
- Water is a precious commodity in some areas
  - Water cost
  - New facilities may need to optimize around water
- Environmental constraints becoming more rigorous
  - Requests for zero liquid discharge are not uncommon now
- Current approach to combustion not rigorous enough
  - Furnaces, flares, combustion devices receive increased scrutiny
  - Present simulation approach may simply call for "duty"
  - Often ignore the "fired side" of the heat supply
    - NOX per unit of heat?
    - VOC per unit of heat?
  - What is the cost of a "unit of improvement"?

PROCESS DEVELOPMENT SYMPOSIUM, JUNE 2-4, 2015, HOUSTON, TX

PDS Process Development Symposium

AIChE
The Global Home of Chemical Engineers

Thank You

Dow Research & Development

PROCESS DEVELOPMENT SYMPOSIUM, JUNE 2-4, 2015, HOUSTON, TX

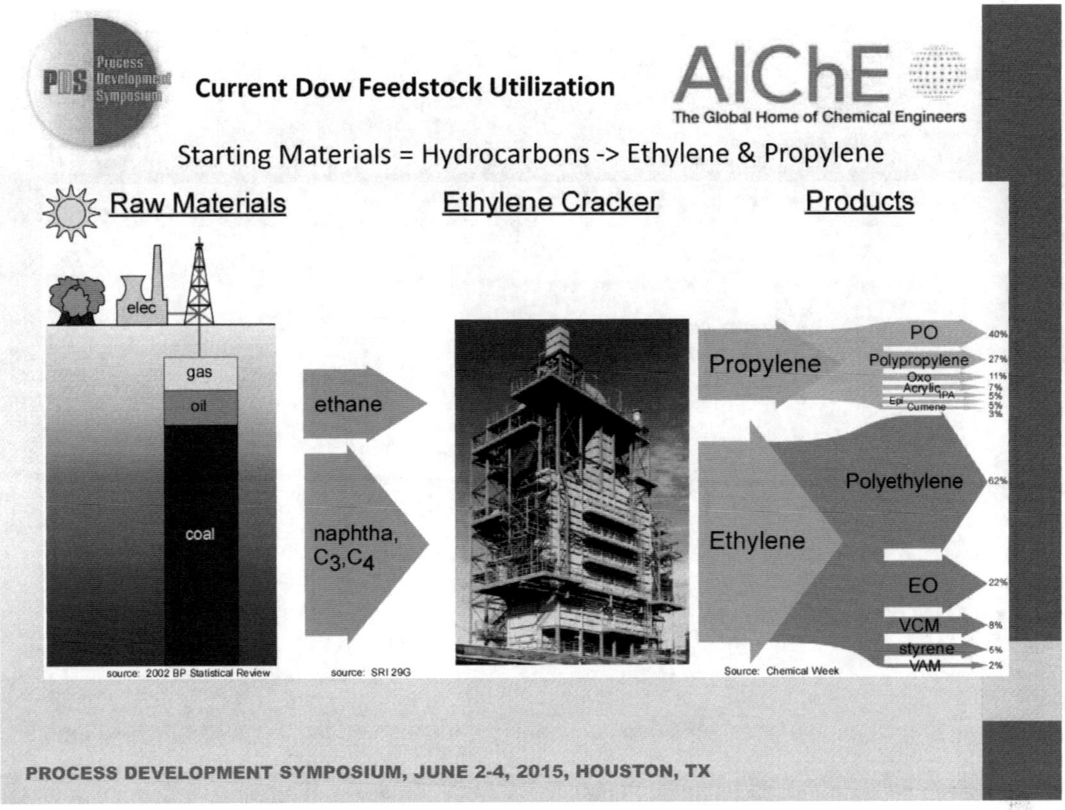

Process Development Symposium
Current Dow Feedstock Utilization
AIChE
The Global Home of Chemical Engineers
Starting Materials = Hydrocarbons -> Ethylene & Propylene
Raw Materials
Ethylene Cracker
Products
elec
gas
oil
coal
ethane
naphtha, C3,C4
Propylene
Ethylene
PO 40%
Polypropylene 27%
Polyethylene 62%
EO 22%
VCM 8%
styrene 5%
VAM 2%
source: 2002 BP Statistical Review
source: SRI 29G
Source: Chemical Week
PROCESS DEVELOPMENT SYMPOSIUM, JUNE 2-4, 2015, HOUSTON, TX

Process Intensification

Opportunities and Challenges for Reaction Engineering

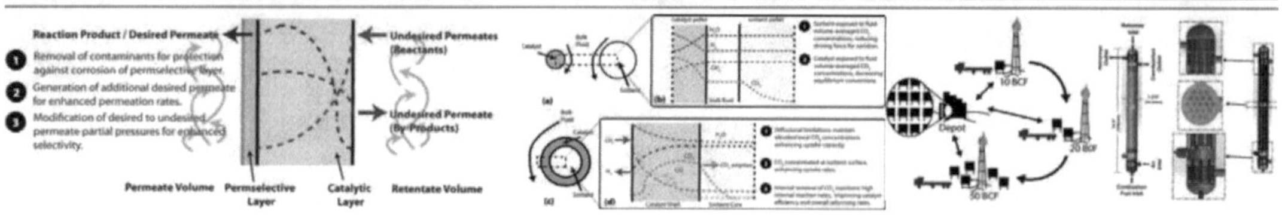

Benjamin A. Wilhite, Associate Professsor
Artie McFerrin Department of Chemical Engineering
Texas A&M University
College Station, TX 77845-3122

TEXAS A&M UNIVERSITY

Presentation Overview

- What is Process Intensification?
- Why Process Intensification?
- Tools and Skills for Process Intensification?
- Mass-Integration in Process Intensification
- Heat-Integration in Process Intensification
- Remaining Challenges: Robustness, Stability of Designs.

TEXAS A&M UNIVERSITY

Motivation: Process Intensification

"...Process Intensification is becoming an immensely important tool to achieve overall sustainability of the chemical process industry and the reduction of our energy footprint through controlled plant size."

- K.T. Deshpande, J. Pendergast, R. Dixit

"...promise spectacular improvements in process plants, markedly shrinking their size and dramatically boosting their efficiency."

- A.I. Stankiewicz, J.A. Moulijn

"Innovations in catalytic reactors...are often the preferred starting point. Over the last two decades, innovative multifunctional reactor systems have been developed to intensify chemical processes by synergistically combining chemical reaction with momentum, heat and mass transport in a single vessel."

- F.M. Dautzenberg, M. Mukherjee

TEXAS A&M UNIVERSITY

Benjamin A. Wilhite June 4th, 2015 – Houston, TX USA 3

Motivation: Process Intensification

"...Process Intensification is becoming an immensely important tool to achieve overall sustainability of the chemical process industry and the reduction of our energy footprint through controlled plant size."

- K.T. Deshpande, J. Pendergast, R. Dixit

"...promise spectacular improvements in process plants, markedly shrinking [process plant] size and dramatically boosting their efficiency."

- A.I. Stankiewicz, J.A. Moulijn

"Innovations in catalytic reactors...are often the preferred starting point. Over the last two decades, innovative multifunctional reactor systems have been developed to intensify chemical processes by synergistically combin[e] chemical reaction with momentum, heat and mass transport in a single vessel."

- F.M. Dautzenberg, M. Mukherjee

TEXAS A&M UNIVERSITY

Benjamin A. Wilhite June 4th, 2015 – Houston, TX USA 4

Process Intensification: Definitions

- The Chemical Reaction Engineering (CRE) Definition of PI:
  - "synergistically combining chemical reaction with momentum, heat and mass transport in a single vessel."
    - Heat-exchanger reactors, coupling endothermic and exothermic reactions in a single unit via direct heat exchanger.
    - Membrane reactors, coupling equilibrium-limited reaction with mass transfer.
- The Process Design Definition of PI:
  - Improving overall thermal, mass efficiency by thermal integration networks of heat exchangers
- The Fine Chemicals / Pharma Definition of PI:
  - Identifying more economical, inherently safer continuous designs and/or processes for fine chemicals and pharmaceutical production to replace existing batchwise-processing.
- The Pragmatic, or "Industry," Definition of PI:
  - Any technology that makes a process significantly more economical.
  - Economical means reduced energy costs, waste....sustainability!

TEXAS A&M UNIVERSITY

Benjamin A. Wilhite June 4th, 2015 – Houston, TX USA 5

Process Intensification: CRE Definition

- The Chemical Reaction Engineering (CRE) Definition of PI:
  - "synergistically combining chemical reaction with momentum, heat and mass transport in a single vessel."
    - Heat-exchanger reactors, coupling endothermic and exothermic reactions in a single unit via direct heat exchanger.
    - Membrane reactors, coupling equilibrium-limited reaction with mass transfer.
- Impact of the CRE Definition of PI:
  - Greater thermal efficiencies, reduced energy waste leads to greater sustainability.
  - Greater mass efficiencies, reduced need for recycle, product separation, leads to reduced mass and energy waste and greater sustainability.

- ***From here on out, this talk will focus on the CRE definition of PI.***

TEXAS A&M UNIVERSITY

Benjamin A. Wilhite June 4th, 2015 – Houston, TX USA 6

Tools Required for PI

- Chemical Reaction Engineering (CRE):
  - Need to understand reactor design fundamentals, catalysis and kinetics of transformations before we can design a mono-functional reactor, let alone a multi-functional one.
- Materials Science and Processing:
  - When coupling transport and kinetics at increasingly small scales, the properties of the materials involved become critical…this is especially the case when designing membrane reactors.
- Computational Tools:
  - By intensifying the reactor via integrating catalytic transformations with heat and mass transport, the need for computational fluid dynamic (CFD) models and optimization algorithms become critical.

TEXAS A&M UNIVERSITY

Benjamin A. Wilhite June 4th, 2015 – Houston, TX USA 7

Two Examples of PI in CRE

- Having established our background and motivation, the remainder of this talk will focus on two areas of research activity in our group at Texas A&M.

1. **Membrane reactors for efficient H_2 recovery from reformate mixtures.**

➢ **Immediate Sustainability Impact**: clean, high-efficiency alternative fuel for urban and suburban transportation (~20% of GHG emissions in USA)

2. **Heat-exchanger microreactors for efficient production of reformate / syn-gas mixtures from natural gas.**

➢ **Immediate Sustainability Impact:** realization of economic alternative to flaring of stranded and/or associated natural gas, and enabling immediate transition to hydrogen fuel economy for transportation (see above).

TEXAS A&M UNIVERSITY

Benjamin A. Wilhite June 4th, 2015 – Houston, TX USA 8

PI in Membrane Reactors

- **Research Goal:** Efficient, simultaneous H_2 production and purification from (i) Reformate mixtures (CO, CO_2, H_2O), (ii) light alcohols (CH_3OH) and/or (iii) hydrocarbons (CH_4).
  - **The Challenge:** Existing hydrogen purification membranes made from Palladium are expensive (low flux per bucks), susceptible to inhibition by CO adsorption, corrosion by alcohols.
  - **The Current Solution**: Design exotic materials or noble-metal alloys to increase permeability, reduce corrosion.

Our solution: Let catalytic reactions enhance performance of permselective materials at minimum complexity, expense by exploiting reaction-diffusion phenomena in catalytic films.

PI in Membrane Reactors:

Catalyst Placement is Critical.

- **CCP Design:** Addition of porous catalyst layer to significantly modify gas composition in contact with underlying permselective layer.

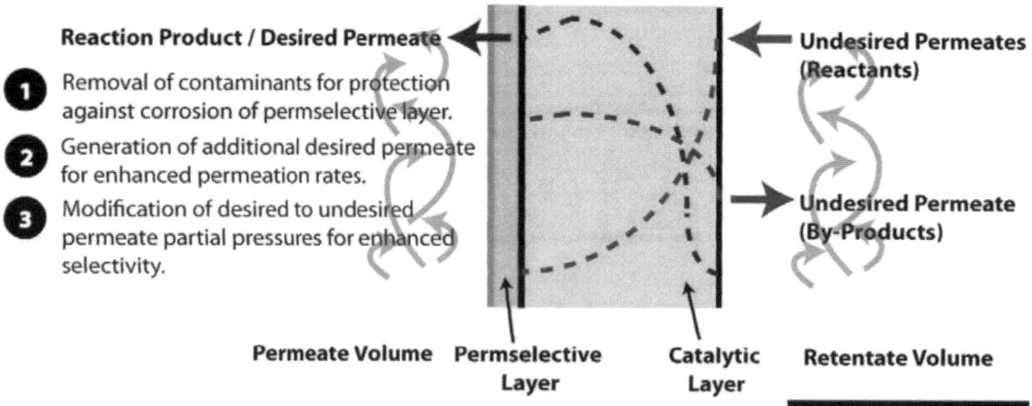

*from Wilhite, *I&EC Res*, 50 (2011), 10185-10193

PI in Membrane Reactors:

- **1st Demonstration of CCP concept:** Enabled first successful H_2 extraction from methanol in a Pd μ-membrane (w/ K.F. Jensen, MIT)

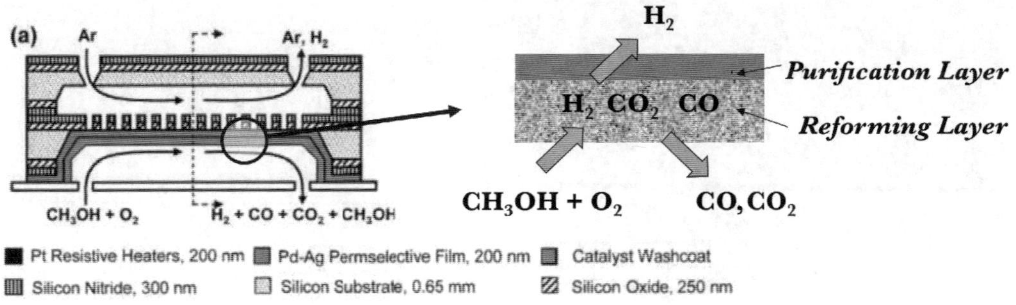

from Wilhite *et al.*, Adv. Mater. 18 (2006), 1701-1704.

PI in Membrane Reactors:

- **Palladium μ-membranes:** Fabricated using Si-MEMS techniques, allowed reduction of active film thicknesses from ~50mm to 200nm.
  - Multiple groups demonstrated order-of-magnitude improvement in purification rates (e.g., MIT, Lehigh, Twente, Karlsruhe)

Original patented architecture developed at M.I.T. by Franz et al. (2002).

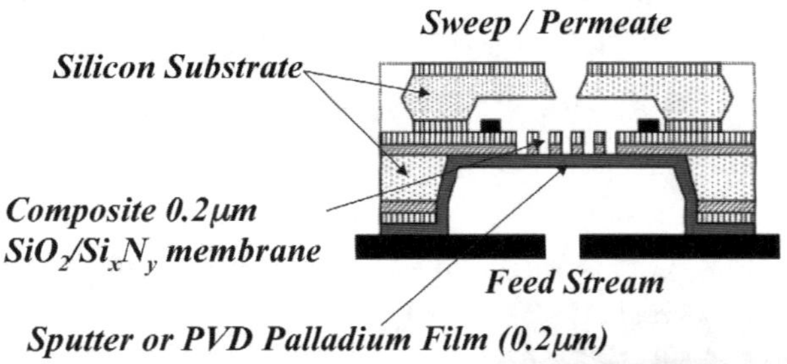

*from Wilhite *et al.*, I&EC Res, 43 (2004), 7083-7087

PI in Membrane Reactors:

- **The Challenge:** Combining chemical reactions with ultra-thin (< 1mm) palladium membrane result in catastrophic degradation.

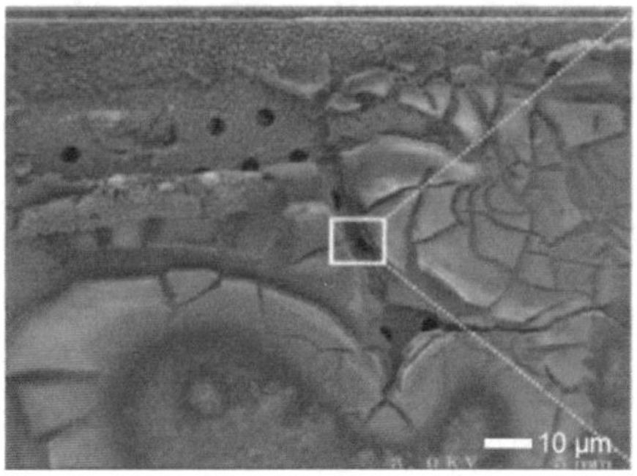

Surface corrosion of 200nm palladium film after exposure to 4% methanol vapor at 350°C for approximately 60 minutes.

Impact: Several publications appeared from 1999 – 2005 reporting ultra-thin Pd membranes, but no successful demonstration of separations outside of ideal conditions (He/H_2; N_2/H_2; Ar/H_2)

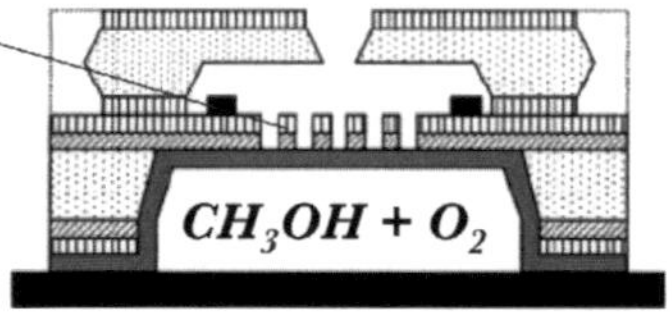

PI in Membrane Reactors

- **The Broader Challenge:** Competitive adsorption of co-reactants (CO especially) block H_2 transport by 10 – 40%, reducing economics of both micro- and conventional Pd-based membranes significantly.
  - *This limits use of Pd membranes at any thickness or any scale!*
  - One of the first thorough, quantitative, models of flux reduction via competitive adsorption reported by Mike Harold's group in 2010 [Israni & Harold, I&EC Res., 2010, 49(21), 10242-10250]
  - Additional data reported by Kim et al., 2010 [Kim et al., I&EC Res., 49(21), 10254-10261]

PI in Membrane Reactors

- **Solution:** Addition of 700mm catalyst layer active for methanol partial oxidation over-top the 200nm Pd layer.

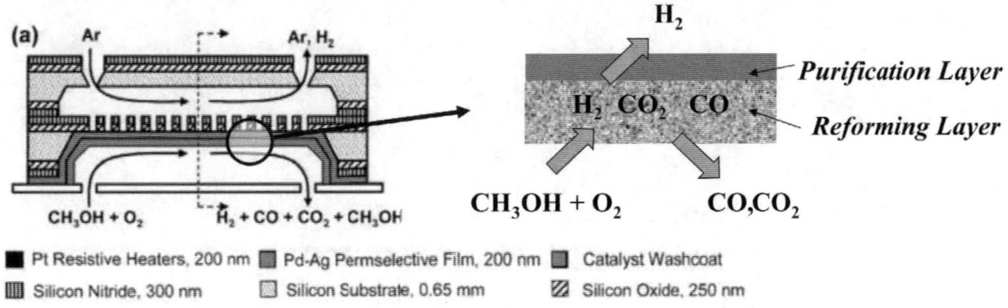

- **Outcome 1**: Hydrogen permeation rates equivalent to that predicted by Sievert's Law assuming 100% conversion of methanol at inlet; i.e. no loss in permeance.
- **Outcome 2**: Stable operation for > 50 hours with no degradation or corrosion.

TEXAS A&M UNIVERSITY

Benjamin A. Wilhite June 4th, 2015 – Houston, TX USA 15

PI in Membrane Reactors

- **CCP Design:** Addition of porous catalyst layer to significantly modify gas composition in contact with underlying permselective layer.

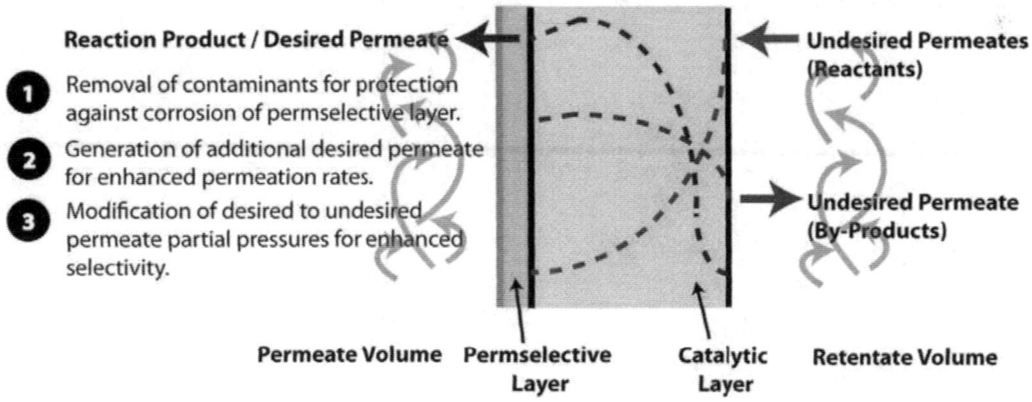

*from Wilhite, *I&EC Res*, 50 (2011), 10185-10193

TEXAS A&M UNIVERSITY

Benjamin A. Wilhite June 4th, 2015 – Houston, TX USA 16

PI in Membrane Reactors

Stage 1: Perform theoretical reaction-diffusion analysis to obtain analytical solutions for identifying initial design rules.

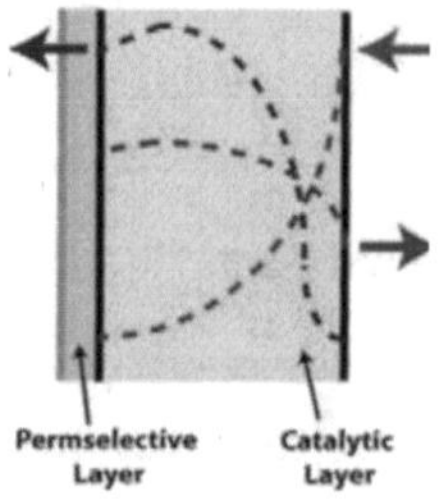

$$\frac{d^2u_i}{ds^2} = \phi^2 \frac{\alpha_{C/i}^{cat}}{\alpha_{C/A}^{cat}} \bullet v_i \cdot r(u_i) \qquad s \in (0,1)$$

$$\frac{du_i}{ds} = \frac{\alpha_{C/i}^{cat}}{\alpha_{C/i}^{mem}} \zeta \cdot u_i \qquad s = 0$$

$$u_i = \theta_i \qquad s = 1$$

Governing Dimensionless Groups

Film Permselectivity	Catalyst Permselectivity	Catalyst:Film Permeance Ratio	Fluid-Phase Stoichiometry
$\alpha_{C/i}^{mem} = \frac{P_C}{P_i}$	$\alpha_{C/i}^{cat} = \frac{D_C}{D_i}$	$\zeta = \frac{P_C}{t} \cdot \frac{RT \cdot L}{D_C}$	$\theta_i = \frac{p_{i,f}}{p_{A,f}}$

TEXAS A&M UNIVERSITY

PI in Membrane Reactors

Stage 1: Perform theoretical reaction-diffusion analysis to obtain analytical solutions for identifying initial design rules.

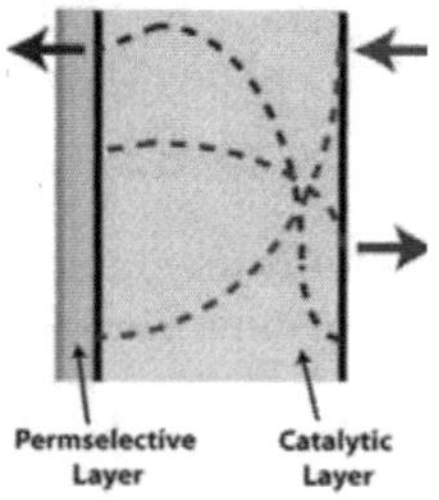

1. **Design Criteria:** z << 1 to prevent excess diffusion resistance placed upon membrane mat'l.
2. **Design Criteria:** F > 7.6 to achieve ~99.9% maximum alteration of composition at membrane surface.
3. **Design Criteria:** K/q_c must be > 1 to avoid reversal of reaction at membrane surface.

Catalyst:Film Permeance Ratio	Fluid-Phase Stoichiometry
$\zeta = \frac{P_C}{t} \cdot \frac{RT \cdot L}{D_C}$	$\theta_i = \frac{p_{i,f}}{p_{A,f}}$

TEXAS A&M UNIVERSITY

PI in Membrane Reactors

Stage 1: Perform theoretical reaction-diffusion analysis to obtain analytical solutions for identifying initial design rules.

Stage 2: Perform detailed design simulations and experiments to demonstrate advantages of CCP in an industrial membrane reactor.

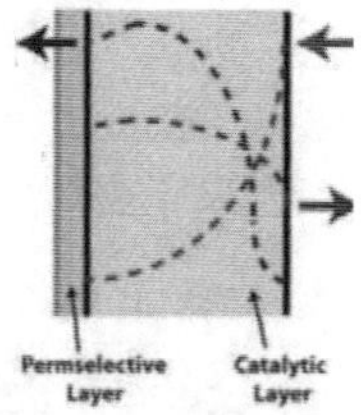

Experiments using 30mm/500mm Pd/cat film

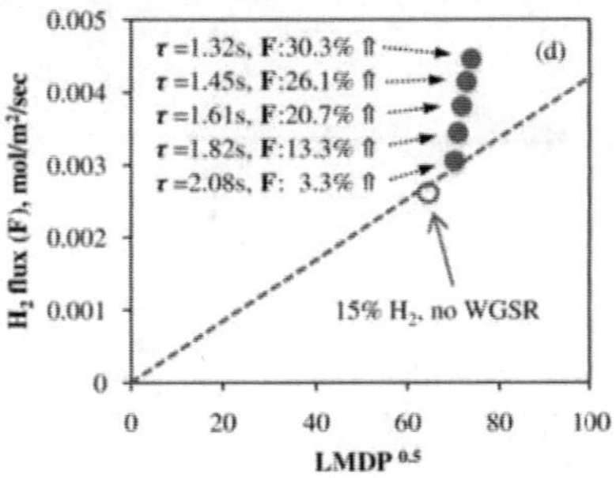

Simulations of 30mm/500mm Pd/cat film

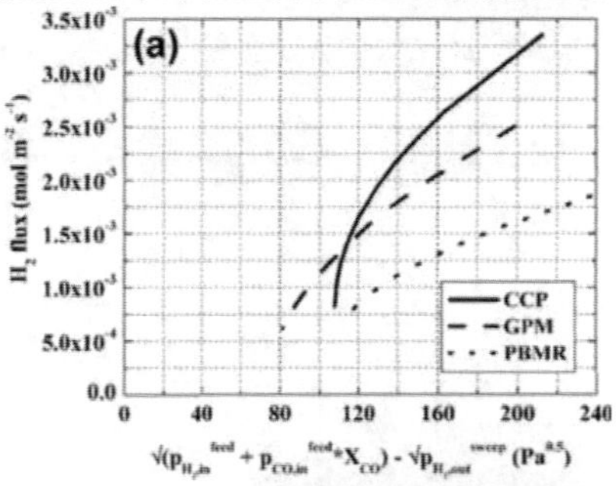

Benjamin A. Wilhite June 4th, 2015 – Houston, TX USA

Application of Design Rules

- **Feedstock:** Assume a typical "heavy hydrocarbon" reformate (28% H_2, 15% H_2O, 15%CO_2, 9%CO, balance N_2).
- **Membrane:** Assume a typical 30-micron thick palladium film.
- **Conditions:** Assume operating temperature of 500°C, compatible with water-gas-shift and Pd membrane.

CCP Suitability?	
Temp.	773 K
K:	20*
$\theta = \theta_C\theta_D / \theta_A$	0.6
K/θ	**34**

Benjamin A. Wilhite June 4th, 2015 – Houston, TX USA

Application of Design Rules

- **Feedstock:** Assume a typical "heavy hydrocarbon" reformate (28% H_2, 15% H_2O, 15%CO_2, 9%CO, balance N_2).
- **Membrane:** Assume a typical 30-micron thick palladium film.
- **Conditions:** Assume operating temperature of 500°C, compatible with water-gas-shift and Pd membrane.

CCP Suitability?	
Temp.	773 K
K:	20*
$\theta = \theta_C\theta_D / \theta_A$	0.6
K/θ	**34**

Catalyst Layer Design	
Temp.	773 K
D_{CO}/RT:	4 x 10^{-10} mol.m^{-1}.s^{-1}.Pa^{-1} **
k:	2.3 x 10^{-6} mol.s^{-1}.m^{-3}.Pa^{-2} *
t_C ($\lambda\phi$ = 7.6)	**470x10^{-6}**

TEXAS A&M UNIVERSITY

Benjamin A. Wilhite June 4th, 2015 – Houston, TX USA 21

Application of Design Rules

- **Feedstock:** Assume a typical "heavy hydrocarbon" reformate (28% H_2, 15% H_2O, 15%CO_2, 9%CO, balance N_2).
- **Membrane:** Assume a typical 30-micron thick palladium film.
- **Conditions:** Assume operating temperature of 500°C, compatible with water-gas-shift and Pd membrane.

CCP Suitability?	
Temp.	773 K
K:	20*
$\theta = \theta_C\theta_D / \theta_A$	0.6
K/θ	**34**

Palladium Layer Analysis	
Temp.	773 K
Permeability:	2 x 10^{-9} mol.m^{-1}.s^{-1}.Pa$^{-0.5}$
t_M	30 x 10^{-6} m
ζ	**0.3 x 10^{-3}**

TEXAS A&M UNIVERSITY

Benjamin A. Wilhite June 4th, 2015 – Houston, TX USA 22

CFD Design Simulations

- **Membrane Reactor Modeling:** Our research group has employed COMSOL simulations to confirm design rules for implementing CCP designs.

Equivalent membrane models of (a,b) CCP membrane, (c) Packed-Bed Membrane Reactor, and (d) stand-alone permselective membrane used to compare CCP against archetypical designs.

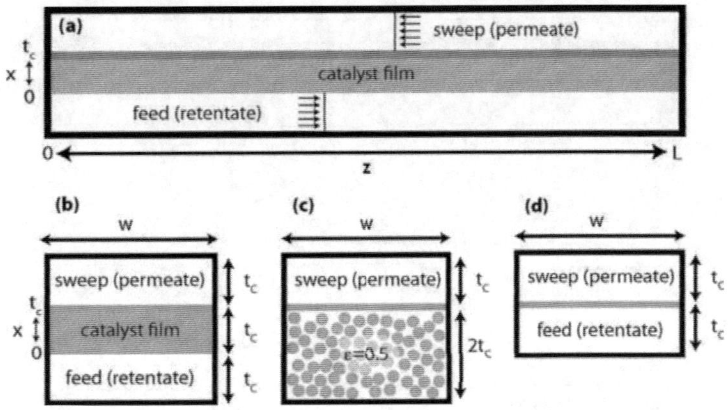

from Romero & Wilhite, Chem. Eng. J., 207-208 (2012), 552-563.

TEXAS A&M UNIVERSITY

CFD Design Simulations

- **Plug-flow** in permeate, retentate volumes
- **Knudsen/Fickian Diffusion & Reaction** in catalyst film.
- **Pd Inhibition** via competitive adsorption, using model of Israni & Harold, I&EC Res, 349 (2010), 380-384.

$$-\frac{1}{A_C^S}\frac{dF_i^S}{dz} = \frac{1}{\alpha_{H_2/i}}\left(\frac{P_{H_2}^{Pd}}{t^{Pd}}\right)\left(\theta_i \cdot \sqrt{p_i^{cat}} - \sqrt{p_i^S}\right)$$

$$\frac{D_{i,eff}}{RT_O}\left(\frac{\partial^2 p_i^{cat}}{\partial x^2} + \frac{\partial^2 p_i^{cat}}{\partial z^2}\right) = R_i$$

$$\frac{1}{A_C^F}\frac{dF_i^F}{dz} = k_{gs}\hat{a}_f\left(p_i^F - p_i^{cat}\right)$$

$$\frac{D_{i,eff}}{RT_O}\frac{\partial p_i^{cat}}{\partial x} = \frac{P_{H_2}^{Pd}}{\alpha_{H_2/i}\cdot t^{Pd}}\left(\theta_i \cdot \sqrt{p_i^{cat}} - \sqrt{p_i^S}\right)$$

$$\frac{D_{i,eff}}{RT_O}\frac{\partial p_i^{cat}}{\partial x} = k_{gs}\left(p_i^{cat} - p_i^{F,S}\right)$$

CFD Design Simulations

- **Simulations:** For simulated diesel reformate (9% CO, 3%CO_2, 15%H_2O, 28%N_2), predicts 30% enhancement in H_2 permeance.

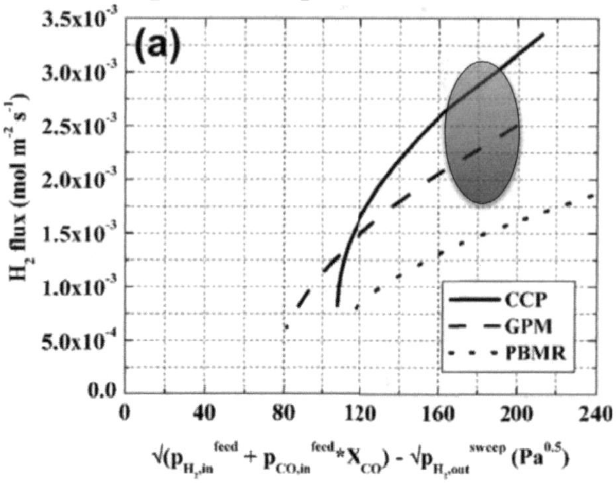

*from Romero & Wilhite, Chem. Eng. J., 207-208 (2012), 552-563.

TEXAS A&M UNIVERSITY

Experimental Validation

- Employing simulated diesel reformate mixture (4.5%CO, 3%CO_2, 7.5%H_2O, 15%H_2), 30-micron electroless-plated Pd membrane and 500 micron CuO-Al_2O_3 catalyst.

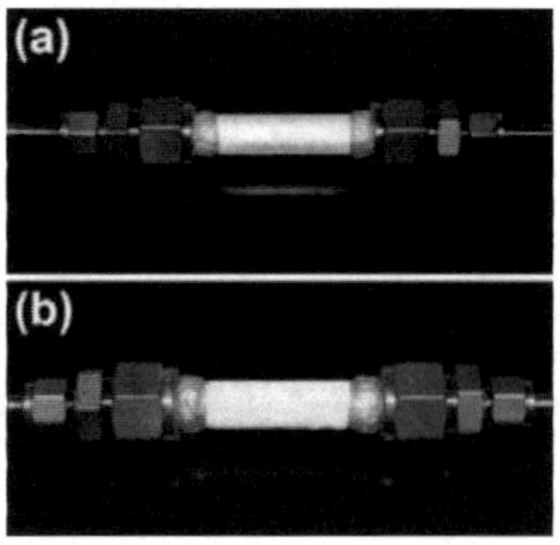

(1) Fabricated 30mm Pd membrane on Mott porous stainless-steel substrate)

(2) Washcoated 500mm CuO/Al_2O_3 catalyst overtop Pd film

*from Kim *et al.*, AIChE Journal, *articles ASAP* (2012)

TEXAS A&M UNIVERSITY

Experimental Validation

- Employing simulated diesel reformate mixture (4.5%CO, 3%CO_2, 7.5%H_2O, 15%H_2), 30-micron electroless-plated Pd membrane and 500 micron CuO-Al_2O_3 catalyst.

(3) Supplied 15%H_2, 7.5% H_2O, 3% CO_2, 4.5% CO to retentate volume, varied feed and sweep flowrates at 350°C.

(4) Observed **up to 30%** increase in overall H_2 flux vs. $LMDP^{1/2}$

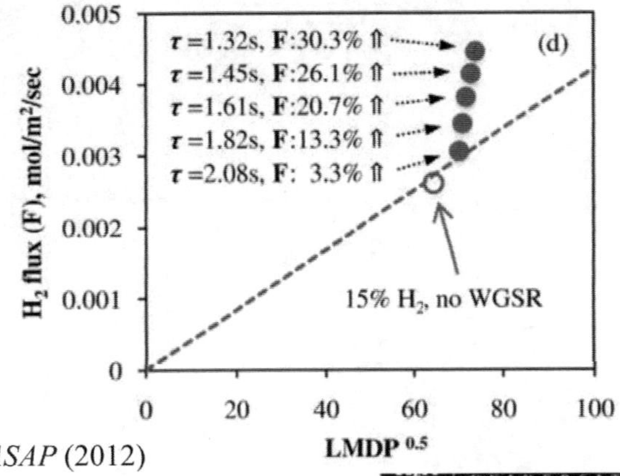

*from Kim *et al.*, AIChE Journal, *articles ASAP* (2012)

Extension to other Rxn-Separations

- **Sorption-Enhanced Reforming Processes**
  - Extension of these same design rules to methane steam reforming coupled with CO_2 sorption indicate equivalent benefits possible.
  - Current SERP designs employ packed-bed admixtures of sorbent (typically CaO-based) and catalyst particles.
  - Preliminary analysis indicates value of employing core-shell (or shell-yolk) designs for maximizing reaction rates, sorbent uptake rates as well as enhancing catalyst lifetime.

Extension to other Rxn-Separations

- **Sorption-Enhanced Reforming Processes**

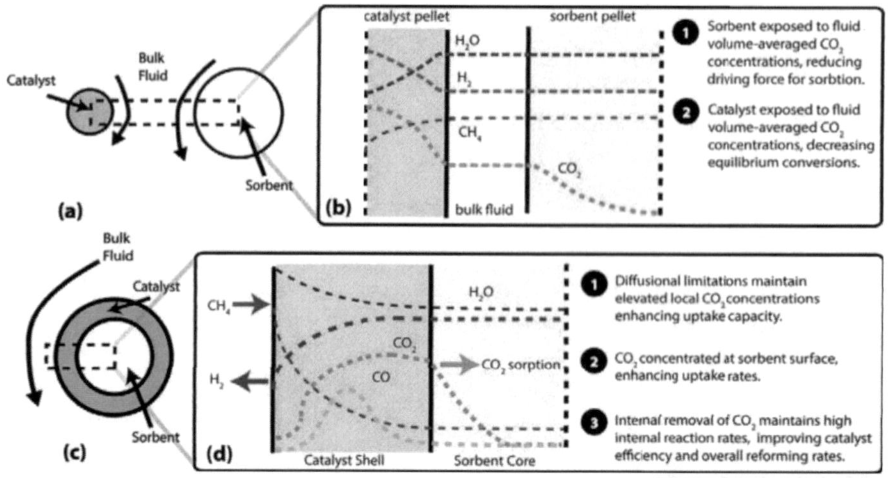

Two Examples of PI in CRE

- Having established our background and motivation, the remainder of this talk will focus on two areas of research activity in our group at Texas A&M.

1. **Membrane reactors for efficient H_2 recovery from reformate mixtures.**

➢ **Immediate Sustainability Impact**: clean, high-efficiency alternative fuel for urban and suburban transportation (~20% of GHG emissions in USA)

2. **Heat-exchanger microreactors for efficient production of reformate / syn-gas mixtures from natural gas.**

➢ **Immediate Sustainability Impact:** realization of economic alternative to flaring of stranded and/or associated natural gas, and enabling immediate transition to hydrogen fuel economy for transportation (see above).

Microreactors: Vision for the Future

- **The Modular Chemical Plant**: Entirely constructed from arrays of interchangeable microchemical "building blocks," allows rapid deconstruction and re-assembly to make different products as needed.
  - Increasingly diverse range of hydrocarbon feedstocks and volatile chemicals market has rendered the traditional "one plant, one process" industry approach obsolete.

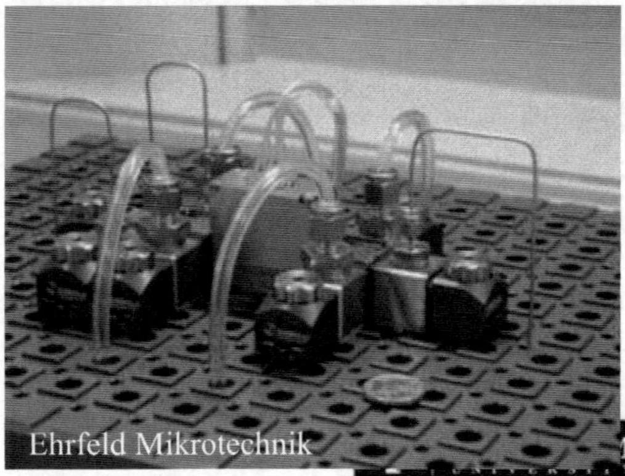

Challenge: robustness or flexibility of components, vs. number of components.

Macroscale vs. Meso/Microscale

Macroscale Chemical Processes

- Length scales of centimeters to meters ($10^{-2} - 10^{2}$)
- System dominated by thermodynamics, equilibrium of overall system
- Transport dictated by inertial forces, "macroscale" effects
- Poor spatial resolution of individual chemical phenomena
- Process chemistry controlled externally via heating coils, agitators, etc.
- Design based upon engineering correlations, 'macroscale' transport and kinetic models.

TEXAS A&M UNIVERSITY

Macroscale vs. Meso/Microscale

Microscale Chemical Processes

- Length scales of microns to millimeters ($10^{-8} – 10^{-3}$)
- Transport dictated by combination of inertia, diffusion, boundary layers
- Process chemistry must be controlled through appropriate structuring of fluid volumes to manipulate local chemistry, physics.
- Requires significant understanding and manipulation of transport, chemical phenomena for success.
- Designs based upon detailed continuum mechanics modeling of momentum, energy and individual chemical transport.

Concept of "Mesoscale Chemical Engineering"

Opportunity to manipulate and control fluid dynamics (i.e., mixing), mass transfer (i.e., diffusion, convection) and heat transfer (convection, conduction) at the sub-millimeter scale.

TEXAS A&M UNIVERSITY

Benjamin A. Wilhite   June 4th, 2015 – Houston, TX USA   33

Industrial Microreactors

- **Microreactor Research (2000's – present):** emphasis has shifted from the novel to the practical. Primarily split between lab-on-a-chip (chemical analysis) and **commercializable energy applications (chemical processing)**.

Srinivasan *et al.*, **AIChE J**, 43 (1997), 3059-3069

Single Microchannel

Holladay *et al.*, **Chem Rev.**, 104(2004), 4767-4790

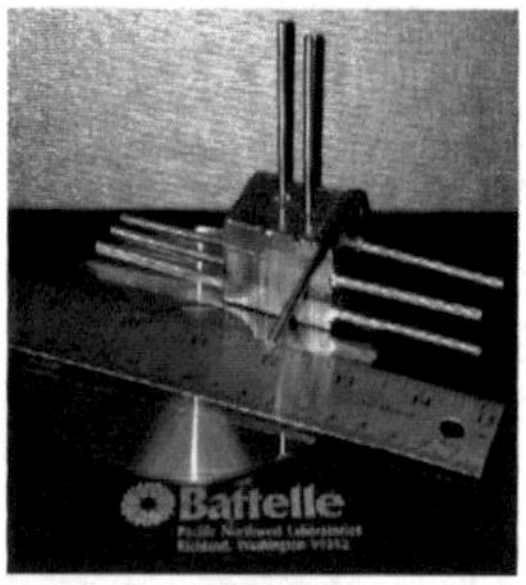

Figure 18. 50 and 100 W integrated fuel processor-cube design.

Microreactor "modules"

Modular reforming processes

TEXAS A&M UNIVERSITY

Benjamin A. Wilhite   June 4th, 2015 – Houston, TX USA   34

Microreactors for Sustainability

- **The Personal "Turn-Key" Chemical Plant:** Fully packaged, turn-key unit capable of converting natural gas to hydrogen as needed (at home or at the local fueling station).
  - Enabling the "Hydrogen Economy" by leveraging existing Natural Gas pipelines requires a turn-key device that can be installed at each fuel station to convert NG to H_2 on-demand.

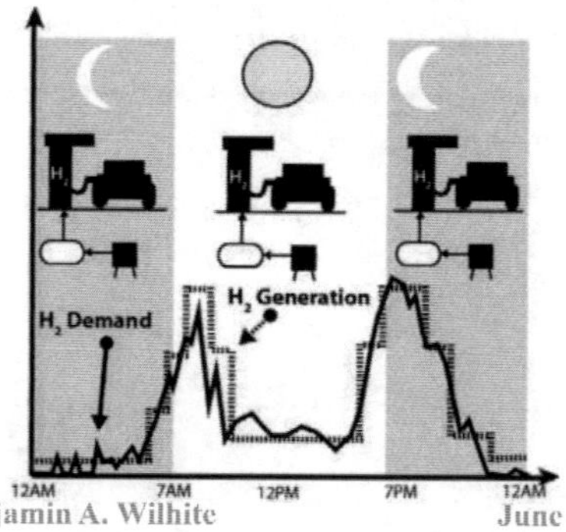

Challenge: how to realize a cost-effective manufacturing design?

Challenge: response times, stability under variable loads.

TEXAS A&M UNIVERSITY

Microreactors for Sustainability

- **The Transportable Chemical Plant:** Constructed from microchemical "building blocks," allows rapid shipping, field-installation and decommissioning of small gas processing facilities for monetizing stranded natural gas.
  - Flaring of associated and/or stranded natural gas is a significant source of greenhouse gas emissions – not to mention wasting valuable non-renewable resources. Need low-cost, transportable processors that can convert gas to liquids (e.g., methanol, diesel)

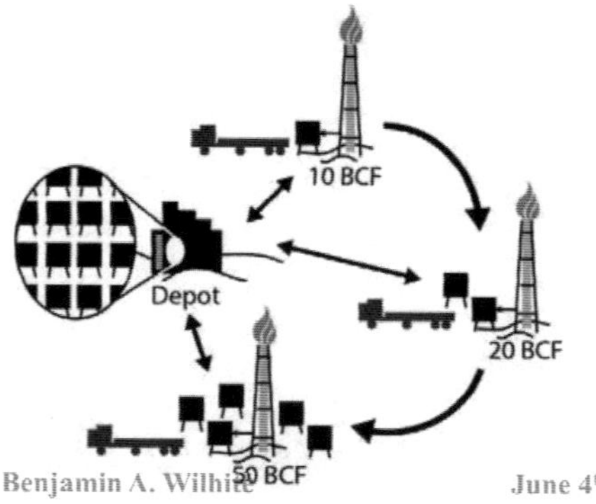

Challenge: maximizing robustness of sub-unit, vs. efficiency of sub-unit.

Do we design system to handle lots of different conditions well, or engineer to handle one condition really well? Trade-off between # of individual components needed and efficiency of system.

TEXAS A&M UNIVERSITY

Microreactors for Sustainability

- In both cases, need to efficiently couple endothermic steam reforming of methane with exothermic combustion for self-powering designs using **Heat-Exchanger Reactor Designs.**

Reformer Volume		Burner Volume
$CH_4 + H_2O \rightarrow CO + H_2$	← Heat Flux	$CH_4 + O_2 \rightarrow CO_2 + H_2O$
$\Delta H = 206$ kJ/mol		$\Delta H = -889$ kJ/mol
$T_{optimal} = 750°C^*$		$T_{optimal} = 750°C^*$

*must not only balance heat duties, but maintain uniform temperatures sufficiently high for reaching equilibrium conversion, yet low enough to prevent catalyst coking or degradation.

Microreactors for Sustainability

- **Challenge:** How to prevent heat losses by conduction, quenching of catalyst films, by controlling/limiting conduction paths?

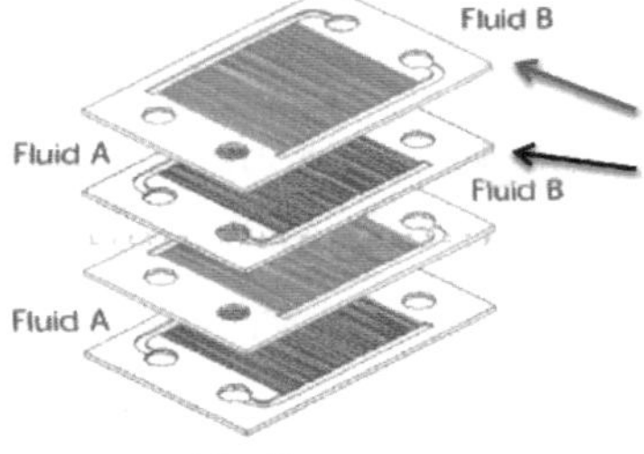

www.ltf-gmbh.com

Challenges for traditional "planar" systems

Endothermic and exothermic catalytic processes exchange heat through solid block – **desired heat pathway.**

But, heat can conduct wherever there is a path – will always choose the "path of least resistance"

Must be able to **isolate heat at the meso-scale,** before it can dissipate on the macro-scale.

Undesired heat pathways

Microreactors for Sustainability

- **Challenge:** How to ensure durable, manufacturable design?

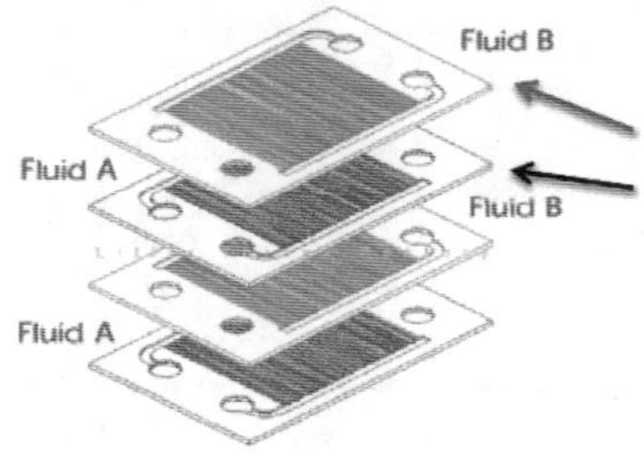

www.ltf-gmbh.com

Challenges for traditional "planar" systems

Each individual plate must be separated machined; can be made cost-effective with metal stamping, batch-wise etching of plates.
But, must seal each plate to the other – adds up to a lot of welding.

Each weld seal must hold up to deformation stresses arising from non-uniform temperatures; seal failure leads to mixing of reforming and combustion chambers.

Microreactors for Sustainability

- **Solution:** The Annular Microreactor (AMR), developed by Power & Energy, Inc.

Power+Energy

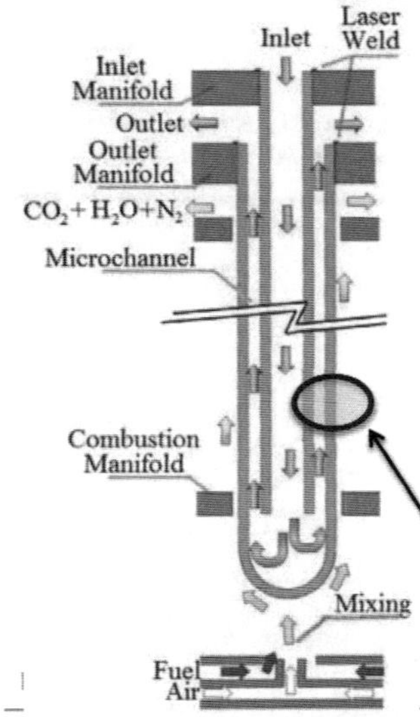

Solution: Inherently Modular, Radial Microreactors

Replacing a single solid block containing multiple microchannels with individual tube-in-tube assemblies – heat cannot conduct from one pair of reaction channels to another: **Mesoscale isolation of reaction heat.**

Radial microreactor concept first proposed in 2012 (Besser, Int. J. H_2 Energy)

Radial microreactors first experimentally demonstrated in 2013 (Bossard et al., J. Power Sources)

Heat transfer occurs only at meso-scale

From Wilhite *et al.*, **Energy & Fuels**, 27(2013), 4403

Microreactors for Sustainability

- To design AMR system, developed detailed computational fluid dynamic model of the microreactor and validated with experimental data [Butcher et al., Int. J. H_2 Energy, 2014, 39(31), 18046-18057].

Experimental Prototype

CFD Geometry

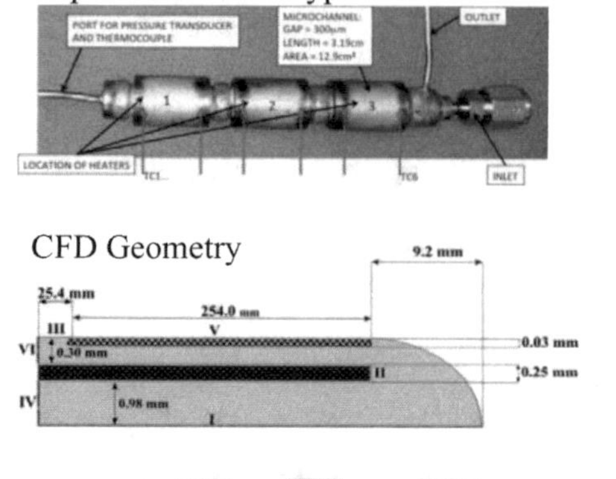

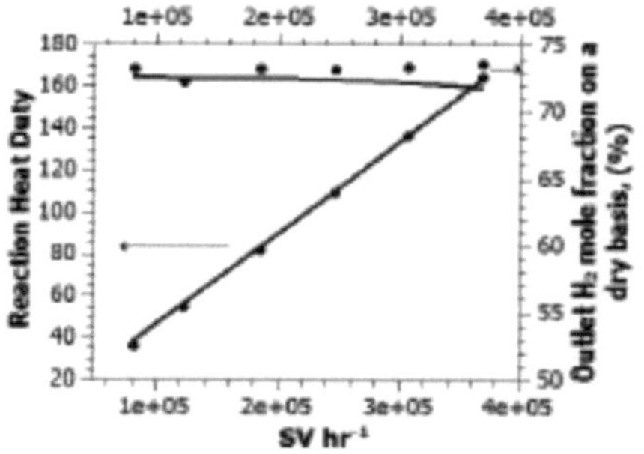

Quantitative match between model and experimental data

Microreactors for Sustainability

- Once validated, CFD model allows identification of nominal operating range of reformer capacities, corresponding heat duties.

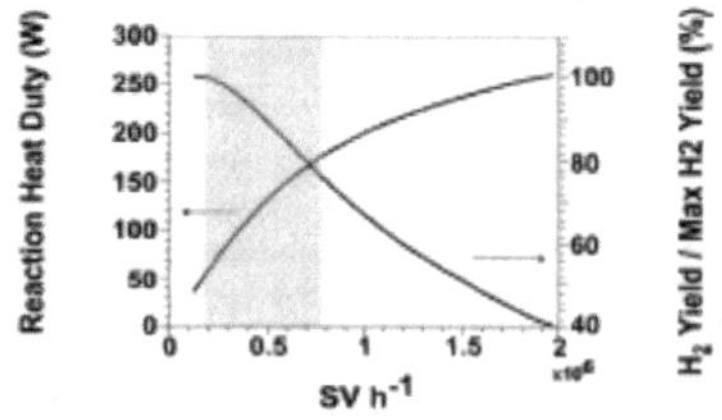

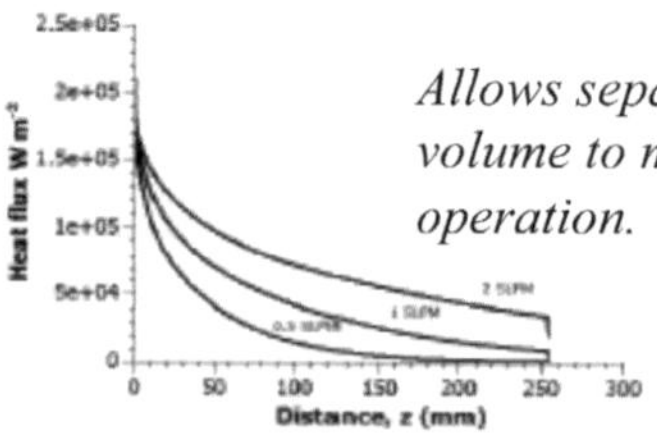

Allows separate design of combustion volume to maintain isothermal operation.

Nominal operating window: 0.5 to 2.0 SLPM methane

- Hydrogen yield 99-75% of theoretical maximum inside operating window
- Thermodynamically unfavorable for Nickel oxidation or Coke formation inside operating window

Heat Duty

- Overall heat duty of 55.8-176.0 W over operating window
- Simple decay style heat duty axially down the reformer
- Only highest flow rate requires significant heat duty at the outlet

Microreactors for Sustainability

- Modeling analysis further enables identification of commercial designs for natural gas monetization and economic analysis of monetization processes using this technology (compared against competiting technologies*).

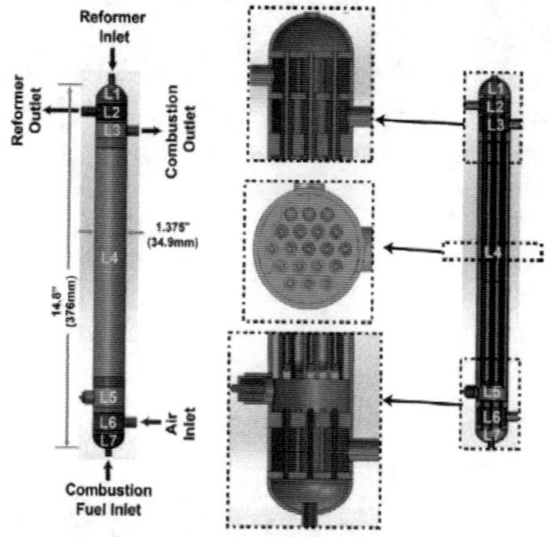

Manufacturing design schematic of 19-AMR cartridge, capable of producing 1-10kW of H_2 from natural gas at overall efficiencies of ~50 – 70%.

Power+Energy

AMR technology developed by Power+Energy, Inc.

*in collaboration with C. Floudas, TAMU

TEXAS A&M UNIVERSITY

Future Challenges

- Robustness of Design:
  - In PI, designs tend to be focused on perfecting a single steady-state. However, in practice these unit operations must be competitive over a range of steady-states.
  - Tradeoff between maximizing efficiency at one steady-state vs. ensuring sufficient efficiencies over range of states.
- Controllability of Designs:
  - In PI, we attempt to achieve a harmonious balance between multiple catalytic and transport phenomena; how stable is this balance?
  - Tradeoff between efficiency and rapid, stable dynamic responses to process fluctuations, startup- and shutdown.

TEXAS A&M UNIVERSITY

Future Challenges

- Need to integrate process instrumentation, control to ensure stable, rapid operation.

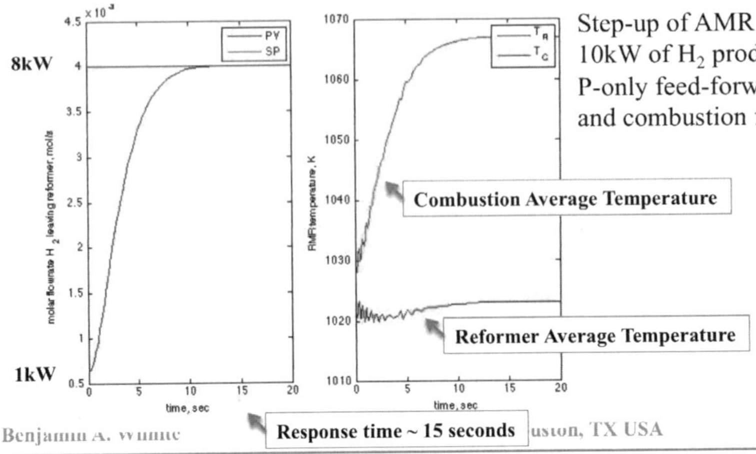

Step-up of AMR system from 1kW to 10kW of H_2 production under dual-loop P-only feed-forward control of reforming and combustion flowrates.

Future Challenges

- Need to integrate process instrumentation, control to ensure stable, rapid operation.

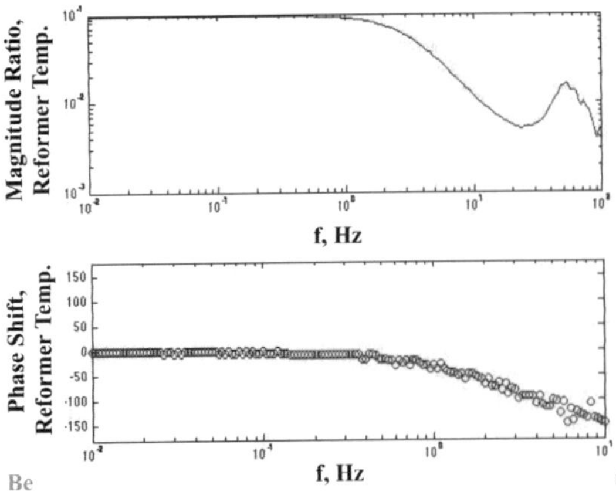

Corresponding Bode Plot for AMR system under dual-loop P-only feed-forward control of reforming and combustion flowrates.

Conclusions

- **Process Intensification**: making chemical transformations more efficient, more portable and more flexible – greater sustainability.
- **Point of Impact:** Natural gas processing has real needs for efficient, flexible conversion of raw materials (NG) into hydrogen, synthesis gas (for subsequent conversion to useful liquids).
- **Skills Needed:** combination of "classical" reaction engineering with materials science, advanced transport analysis.
- **Membrane Reactors:** must balance reaction rates with transport rates, but must also consider the configuration / architecture.
- **Microreactors:** must balance complex designs with manufacturable designs, and consider packaging, durability.
- **Remaining Challenges:** transition from focusing upon a single, perfect steady-state; must account for robustness and controllability.

TEXAS A&M UNIVERSITY

Benjamin A. Wilhite   June 4th, 2015 – Houston, TX USA   47

Acknowledgements

Our Research Group:

Holly Butcher (Ph.D., 2016)
PVK Srikanth (Ph.D., 2016)
Xiahong Cui (Ph.D., 2015)
Elva Romero (Ph.D., 2015)
Haomiao Zhang (Ph.D., 2015)
Dr. Daejin Kim, (Ph.D. 2010), ORNL
Dr. Bhanu Kuncharam (Ph.D. 2014); WPI
Mr. Kevin Barnett (B.S., 2013); U-Wisc.
Dr. Aravind Suresh (Ph.D. 2011), U-Conn.
Dr. Angela Moreno (Ph.D. 2010), 3M

Collaborators:

C. Barry Carter (UCONN)
T. Cagin, J. Grunlan (TAMU)
P. Bossard, C. Quenzel (Power+Energy, Inc)
X. Wu (Eastman Chemicals)

Funding:

National Science Foundation
Office of Naval Research
DuPont Young Investigator Program
Eastman Chemical Coorporation
Power+Energy, Inc.
ACS Petroleum Research Fund
Mary Kay O'Connor Safety Center (TAMU)

Power+Energy Fueling the Hydrogen Economy

AMR technology developed by Power+Energy, Inc.

TEXAS A&M UNIVERSITY

Benjamin A. Wilhite   June 4th, 2015 – Houston, TX USA   48

Thank You!

Wilhite Research Group:
Chemical Reaction Engineering at TAMU
http://research.che.tamu.edu/groups/wilhite/wp_home/

TEXAS A&M UNIVERSITY

Benjamin A. Wilhite June 4th, 2015 – Houston, TX USA 49

Source Material, or Suggested Readings

- H. Butcher *et al., Int. J. Hydrogen Energy,* **39** (2014), 18046-18057.
- B.A. Wilhite *et al., Energy Fuels,* **27**(2013), 4403–4410.
- S. Damodharam, B. Kuncharam and B.A. Wilhite, *Energy Fuels*, **27** (2013), 4411–4422.
- D. Kim, K. Barnett and B.A. Wilhite, *AIChE J.,***59** (2013), 1627.
- E. Lugo and B.A. Wilhite, *Chem. Eng. J., 207-208(2012), 552-563.*
- B.A. Wilhite, *Industrial & Engineering Chemistry Research, 50(17), 10185-10193 (2011).*
- E.L. Romero and B.A. Wilhite, *Chemical Engineering Journal, 175, 433-442 (2011).*
- G.S. Honda and B.A. Wilhite, *Chemical Engineering Science, 68(1), 663-667 (2011).*
- A. Moreno, S. Damodharan and B.A. Wilhite, *I&EC Research,* **49,** 10956-1094 (2010).
- D. Kim *et al., I&EC Research,* **49**, 10254-10261 (2010).
- A. Moreno and B.A. Wilhite, *Journal of Power Sources,* **195(7)**, 1964-1970 (2010).
- D. Kim *et al., J. Mem. Sci.,* **340**, 109-116 (2009).
- A. Moreno, K. Murphy and B.A. Wilhite, *Ind. Eng. Chem. Res.,* **47**, 9040 (2008).
- B.A. Wilhite *et al., Advanced Materials*, **18**, 1701 (2006).
- B.A. Wilhite, M.A. Schmidt and K.F. Jensen, *Ind. Eng. Chem. Res.*, **43**, 7083 (2004).

TEXAS A&M UNIVERSITY

Benjamin A. Wilhite June 4th, 2015 – Houston, TX USA 50

Roger A. Harris
AIChE Process Development Symposium, Houston, June 2015

Commercializing and deploying microchannel FT reactors for smaller scale GTL facilities

VELOCYS

Contents

- Who is Velocys?
  - — Velocys project pipeline
- Our technology at a glance
  - — Microchannel reactors
  - — Catalyst
- The need for piloting within the lifecycle of a novel technology
  - — The first steps are taken in the lab
  - — Our piloting facilities
    - Ideal for scale-up
    - Intensifying our offering
    - Addressing transient conditions
    - Catalyst handling
- ENVIA Energy OKC update

Velocys
The company at the forefront of smaller scale GTL

- **Leader** in smaller scale gas-to-liquids technology
  - 15 years and >$300 million invested in product development
  - Exhaustive global patent protection (>7,600 granted GTL patent claims)
- First class **partners** offering a **complete GTL solution**
  - Haldor Topsøe, Ventech, Hatch, Mourik, SGS, Shiloh
- **Commercial roll-out underway**

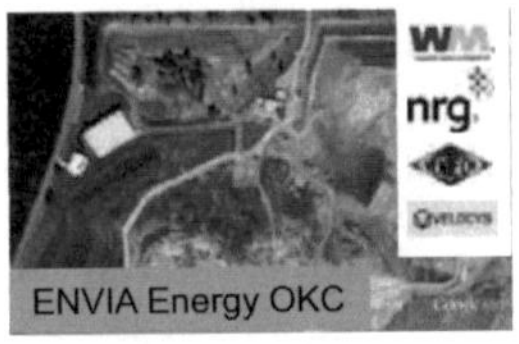

- **Well capitalised** with **strong resources**
  - Commercial center in Houston, Texas; technical centers near Columbus, Ohio and Oxford, UK
  - Permanent pilot plant in operation

VELOCYS 3

Other public-domain commercial projects announced
Pipeline growing and progressing

- 4,800 bpd plant supplied from Marcellus; future expansion planned to 10,000 bpd
- Initial engineering complete (summer 2014); permits received (Jan & June 2014)
- FID targeted end 2015

Red Rock Biofuels

- 1,100 bpd biomass-to-liquids plant in Oregon, USA
- FEED study complete (summer 2014)
- Awarded **$70 million construction grant** from DOD (Sep 2014)
- Targeting FID **H2 2015**

- 2,500 bpd waste-to-liquids plant at former Coryton oil refinery in Thurrock, Essex, UK
- Pre-FEED complete (summer 2014), now **entering FEED**
- BA signed offtake for all jet fuel from facility (worth $500m)

VELOCYS 4

Microchannel reactors
Backbone of Velocys FT technology

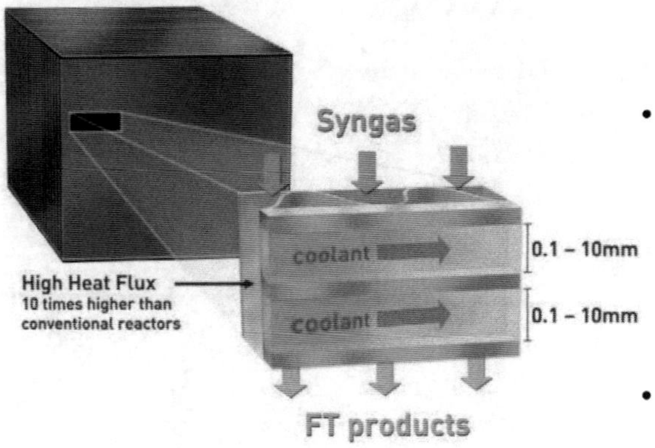

- Microchannels enhance heat and mass transfer rates
- Alternating process & coolant layers allow close coupling of exothermic reaction with steam generation aiding the robust performance
- Particulate catalyst in small process channels

Origin of exceptional catalyst performance
Patented organic matrix combustion (OMX) method

- **Traditional catalyst manufacturing**
  - — Produces broad distribution of cobalt particle sizes
- **OMX catalyst manufacturing**
  - — Produces optimized smaller and more uniform cobalt particle sizes
  - — **Higher activity** from smaller particles
  - — **Greater stability** from narrower particle size distribution

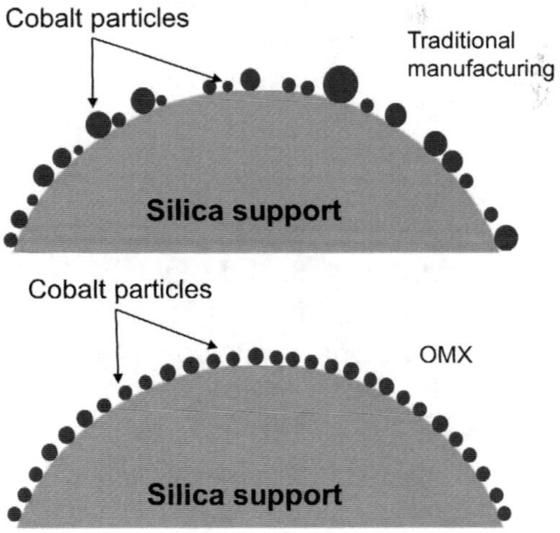

Single channel reactor
Laboratory testing tool

- 1 process channel
  - No catalyst dilution
  - Coolant channels with hot oil circulation
- Validate commercialization requirements
  - Stability
  - Regenerability
  - Catalyst life
  - Predictive models
  - Tolerance to feed contaminants, upsets

VELOCYS 7

Pilot plant and training facility

- Integrated GTL pilot plant in Ohio
  - Process
  - Catalyst handling
  - Mechanical and component testing
- Provides
  - Performance data to support differing client designs
  - Product for client studies
  - Permanent training facility for plant operators
- Platform for
  - Developing our own field support staff
  - Demonstrating future product generations

Pilot plant

VELOCYS 8

Verified performance and scale up methodology

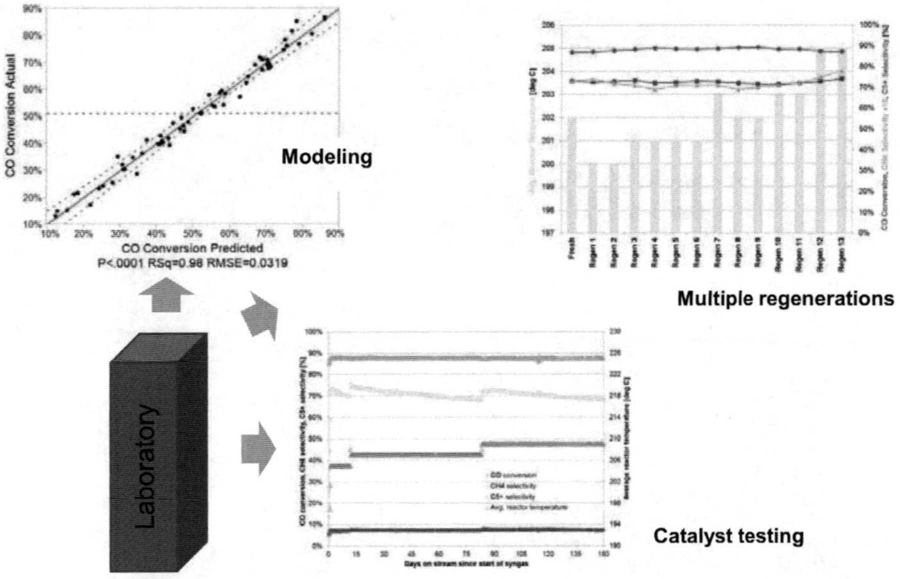

VELOCYS 9

Verified performance and scale up methodology

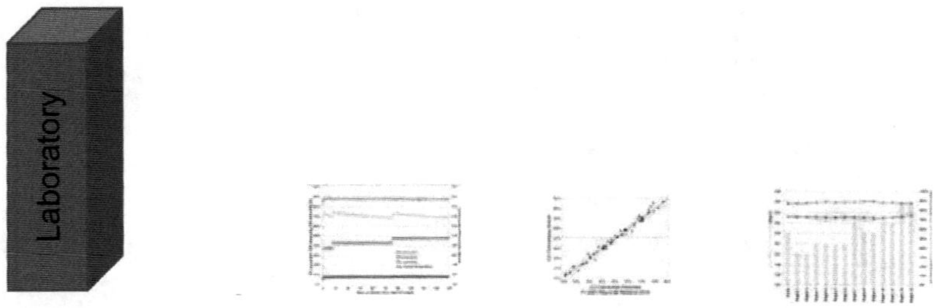

VELOCYS 10

Verified performance and scale up methodology

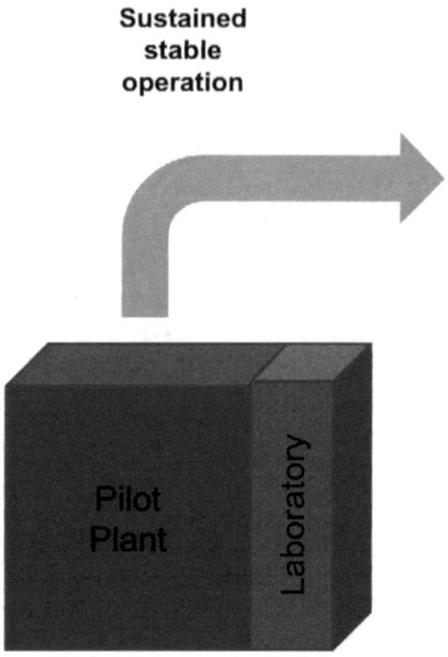

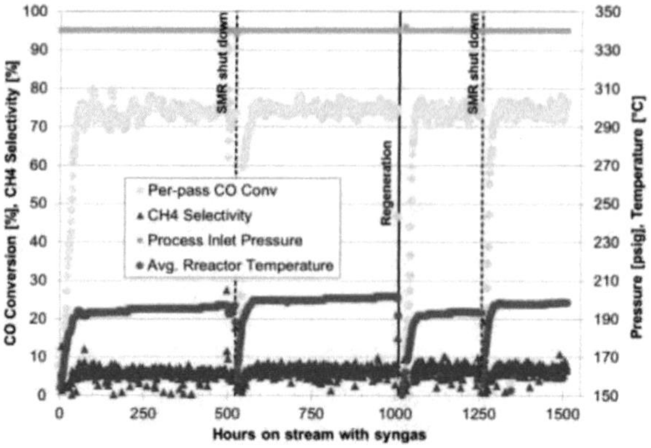

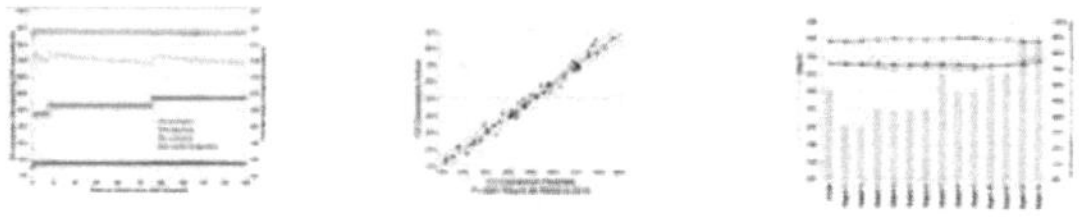

Verified performance and scale up methodology

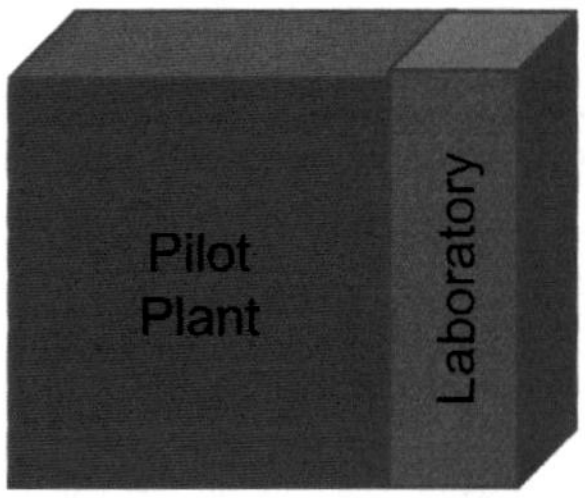

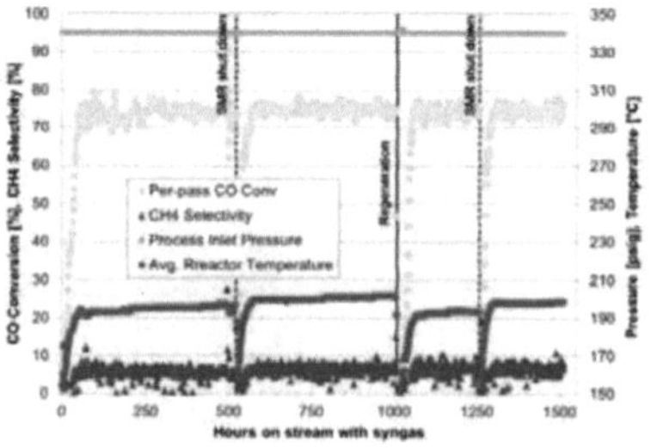

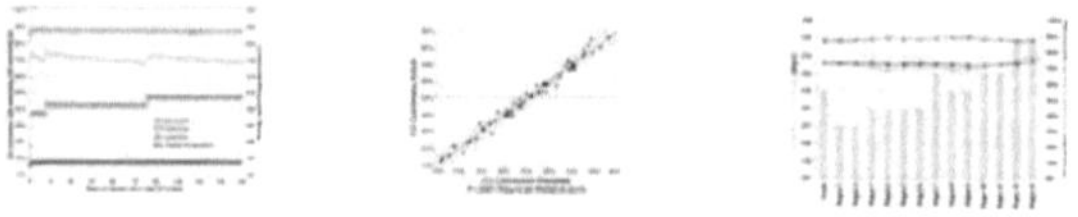

Verified performance and scale up methodology

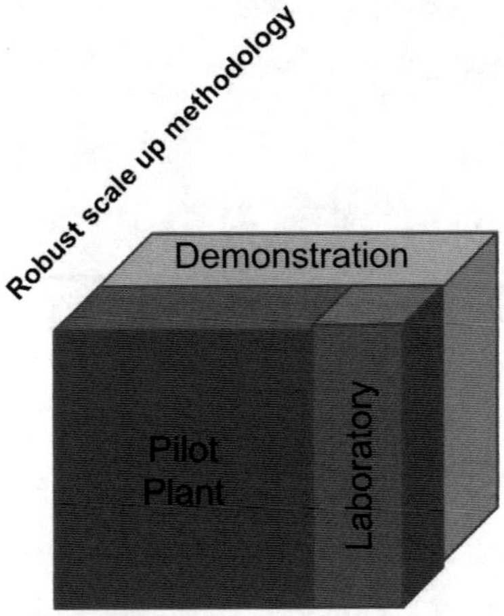

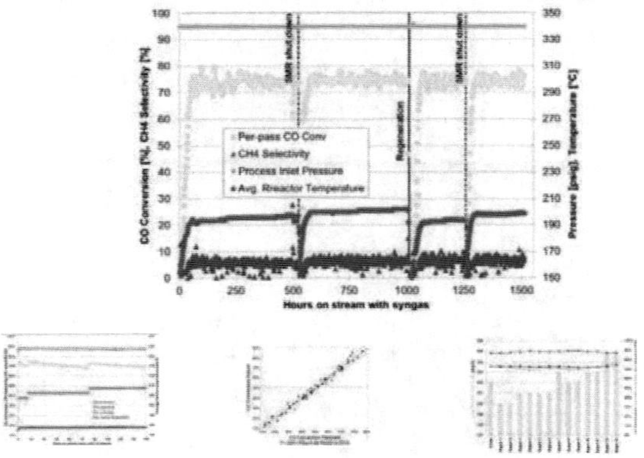

VELOCYS 13

Verified performance and scale up methodology

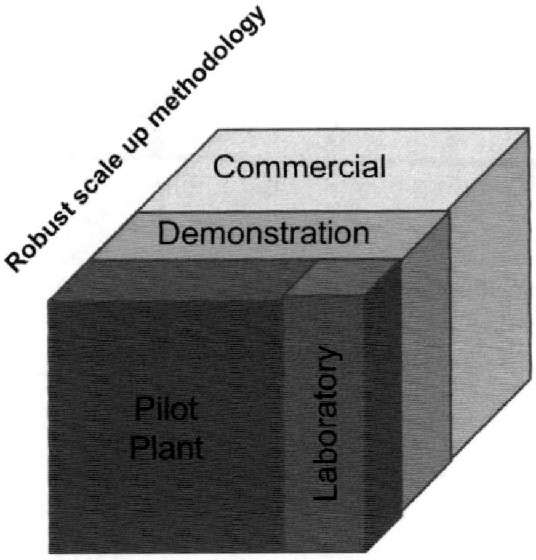

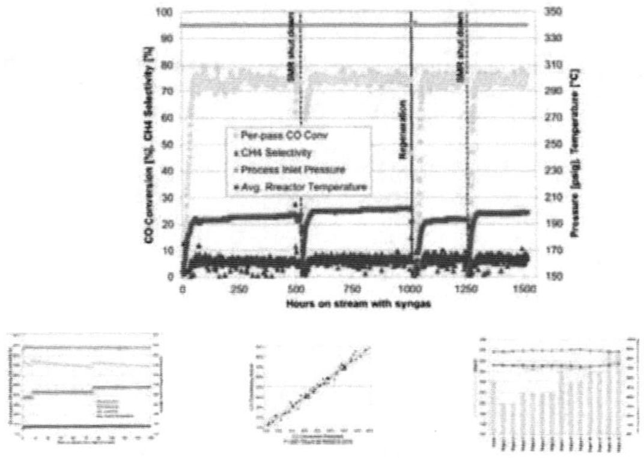

VELOCYS 14

Verified performance and scale up methodology

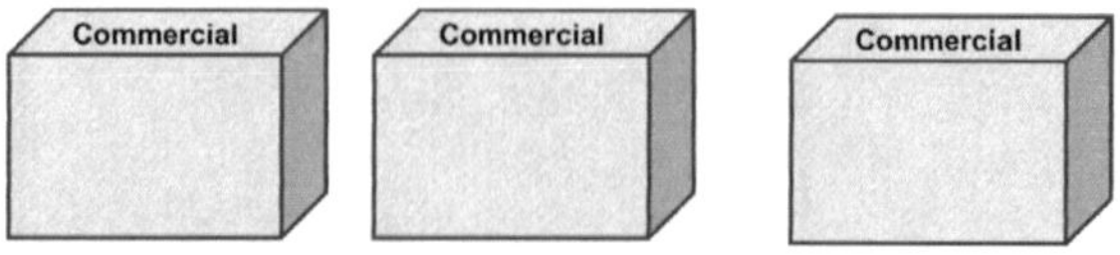

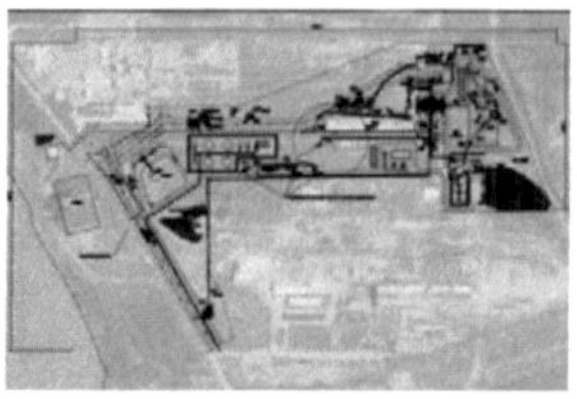

Duplicate reactors and trains for commercial facilities

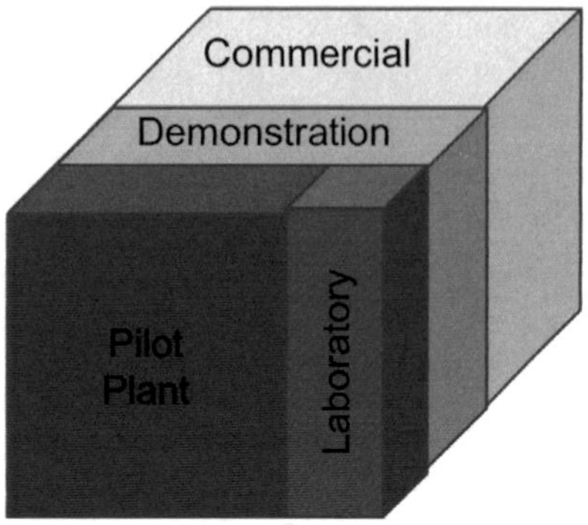

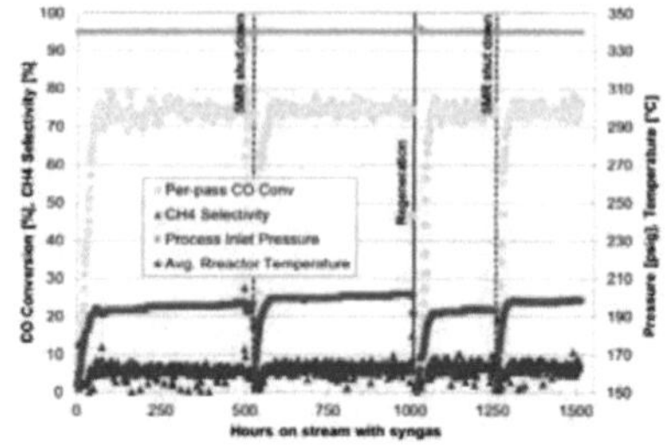

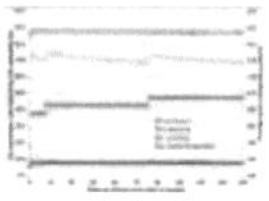

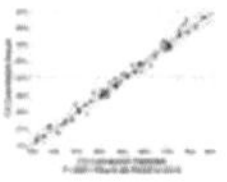

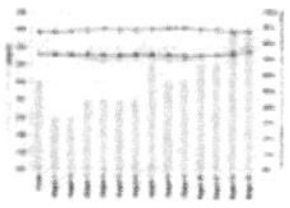

Pilot plant runs over the last 18 months
Process intensification

- Successful piloting of commercially produced catalysts
- Continuous improvement without changing our technology basis
- Process intensification results in significant capital cost improvement

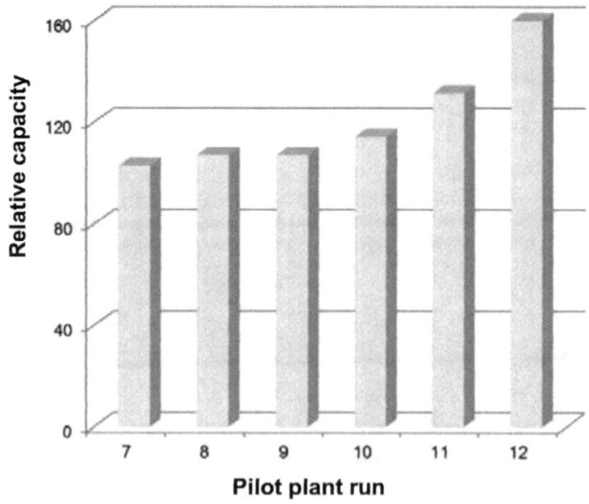

Piloting alternative scenarios
Expect the unexpected

- Commercial plants in any industry are subject to feed and situational variances
- A robust technology needs to be able to be resilient to these changes
- Piloting provides an excellent way to validate dynamic tools and understanding as well as to create confidence in the technology's ability to perform in various operational modes
- Approximately 1,000 hours dedicated to alternative modes and regimes this year

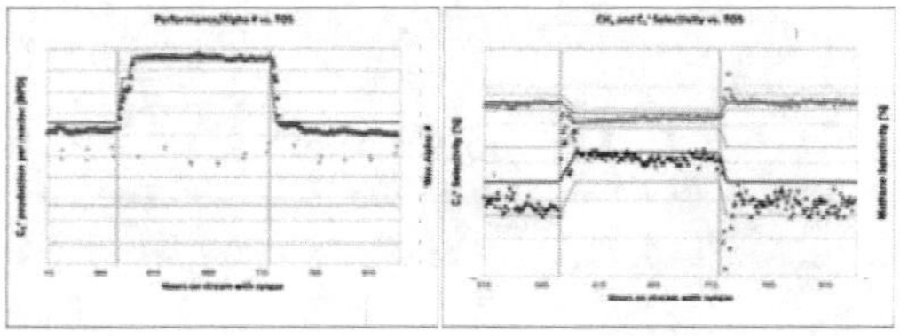

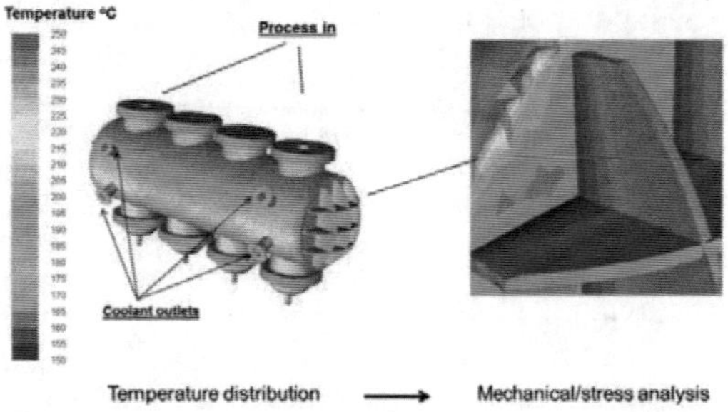

Catalyst handling

- Partnering and collaboration with a strong world leader in catalyst handling to enable success

MOURIK

"Mourik is one of the few companies in the world that have mastered this task. We don't want to be the biggest, but we do want to be the best. And we translate this ambition into a constant flow of innovations that help our clients improve their performances in a sustainable way."

- Good catalyst handling demands:
  - Safety
  - Efficiency
  - Uniformity
- This area is very suitable to continuous improvement and innovation
  - ~ 10+ patents and growing
- We have developed semi-automated loading and unloading systems that have been trialed at various stages including commercial cores

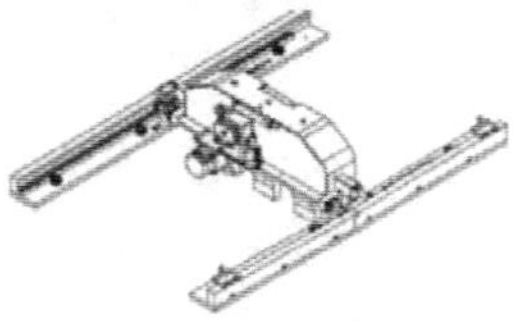

- We have performed a pilot plant catalyst turn around using our commercial methods with gas-out / gas-in times comparable to commercial expectations

Scale of catalyst handling tests and demonstrations

Laboratory

Multiple tests and experiments to determine and optimise

- Packing density parameters used in statistical and flow modelling
- Procedure for preparing catalyst for discharge
- Impact of deviations from normal operation

Pilot Plant

- Demonstrated efficacy of our procedures and methods for loading and discharge
- Measured performance and validated models
- A pilot plant reactor was operated for a complete run, prepared for discharge, discharged, re-loaded and restarted successfully – all within commercially acceptable time constraints

Commercial

- A full scale (3-core) commercial unit was used to develop and train personnel on catalyst loading procedures
- A single core commercial variant is used to improve and innovate new catalyst handling techniques
- Commercially produced cores are used for extensive long term life cycle tests

ENVIA Energy
Oklahoma City project

- **A landmark for GTL**
  - Landfill biogas + natural gas
  - Major companies committing to smaller scale GTL
- **Project on schedule**
  - Ground breaking 15th May 2015
  - Modules currently being constructed
  - Commissioning and start-up H1 2016
- Will be a commercial reference plant for Velocys – a **major milestone**
  - Demonstrate parallel operation of full scale Velocys reactors
- JV identifying sites for the next project(s)
  - **Strategic differentiator** for WM

WM WASTE MANAGEMENT

nrg

VELOCYS

VENTECH

Houston-area companies break ground on landfill gas-to-diesel project

Posted on May 15, 2015 | By Rhiannon Meyers

Work has started to prepare a site in Oklahoma City for a new gas-to-liquids plant that will convert landfill gas into petroleum products. (Envia Energy)

Module and vessel fabrication underway

At Ventech (Pasadena, Texas)

Thank you

Roger A. Harris
GTL Technology Manager
roger.harris@velocys.com

Plain City, Ohio office
+1 614 733 3300

Houston office
+1 713 275 5840

www.velocys.com

Acknowledgements

VELOCYS 23

Think Smaller

VELOCYS

EASTMAN

Mini-plant Demonstration Integrated with Process Development and Final Commercial Design

Bob Huss and Dan Terrill
June 5, 2015

PROCESS DEVELOPMENT SYMPOSIUM, JUNE 2-4, 2015, HOUSTON, TX

PDS Process Development Symposium

Outline

- Background
- 5-Stage scale up strategy
- Stage 3: Bench process innovation
- Stage 4: Miniplant demonstration
- Preliminary commercial design
- Summary

PROCESS DEVELOPMENT SYMPOSIUM, JUNE 2-4, 2015, HOUSTON, TX EASTMAN

Background:

2-step chemistry:

$$D + E \leftrightarrow FG \rightarrow C$$

PROCESS DEVELOPMENT SYMPOSIUM, JUNE 2-4, 2015, HOUSTON, TX EASTMAN

Technology Scale-up Strategy

1. Concept on paper
2. Batch lab demonstration
   Chemistry, reaction rate, selectivity, batch separations
3. Continuous stand-alone lab demonstration
   Optimization of rate/selectivity, catalyst life, separation recoveries
4. Integrated process (mini-plant) demonstration
   Smallest-possible-scale, continuous process using all unit operations closing all recycles at commercial-design conditions and compositions, typically 5-15 ml/min. Evaluation of recycled, minor-component effects and discovering unknown unknowns
   a. Semiworks demonstration, if necessary
      Producing salable product for customer qualification.
      Matching commercial compositions and closing all recycles.
5. Full-scale production

PROCESS DEVELOPMENT SYMPOSIUM, JUNE 2-4, 2015, HOUSTON, TX EASTMAN

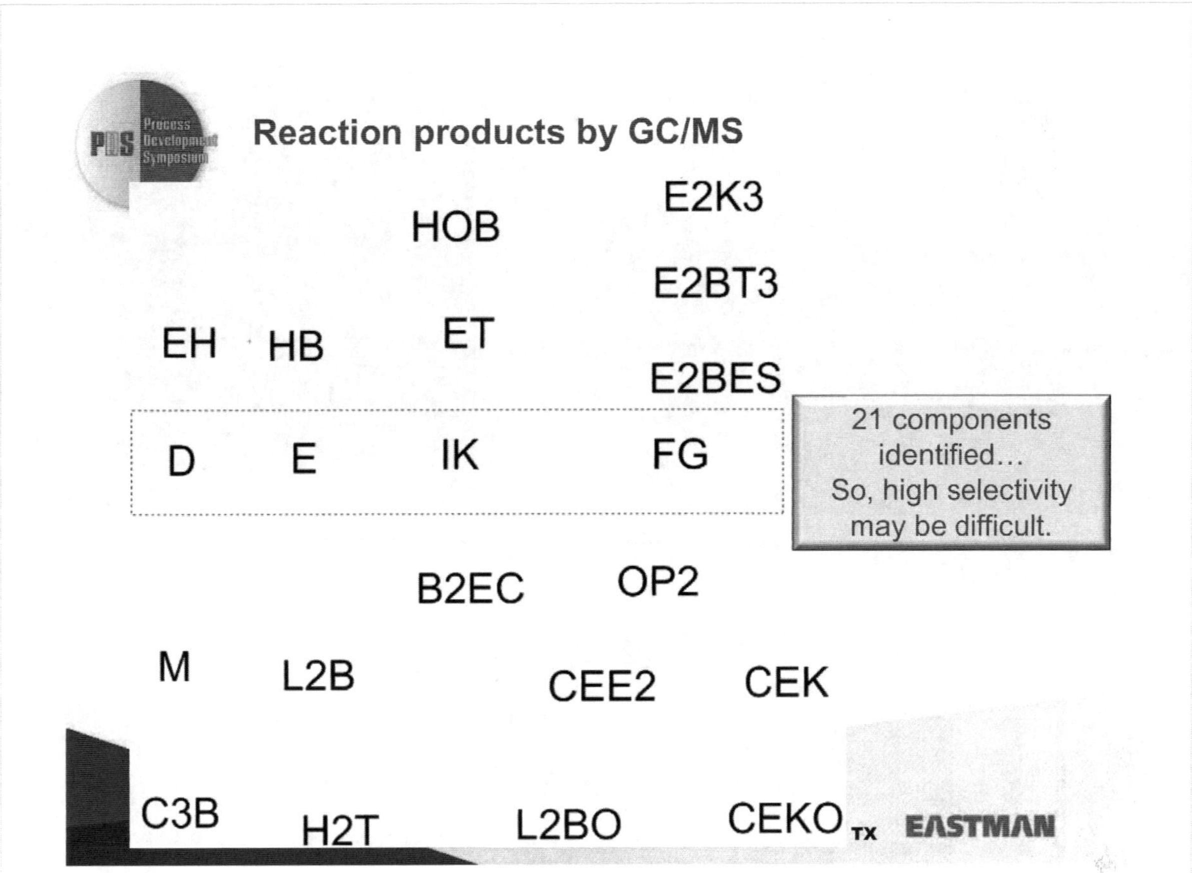
PDS Process Development Symposium
Reaction products by GC/MS
HOB
E2K3
E2BT3
EH
HB
ET
E2BES
D
E
IK
FG
21 components identified…
So, high selectivity may be difficult.
B2EC
OP2
M
L2B
CEE2
CEK
C3B
H2T
L2BO
CEKO
TX
EASTMAN

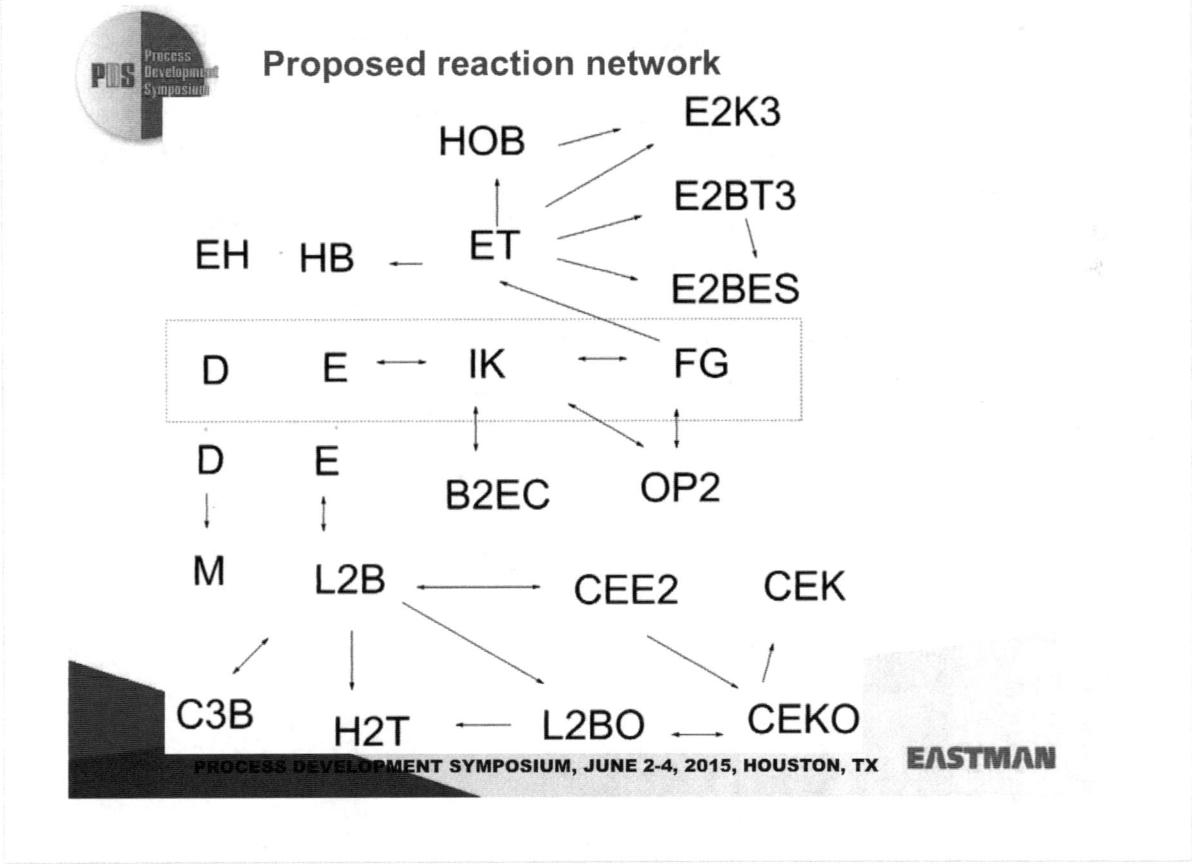
PDS Process Development Symposium
Proposed reaction network
HOB
E2K3
E2BT3
EH
HB
ET
E2BES
D
E
IK
FG
D
E
B2EC
OP2
M
L2B
CEE2
CEK
C3B
H2T
L2BO
CEKO
PROCESS DEVELOPMENT SYMPOSIUM, JUNE 2-4, 2015, HOUSTON, TX
EASTMAN

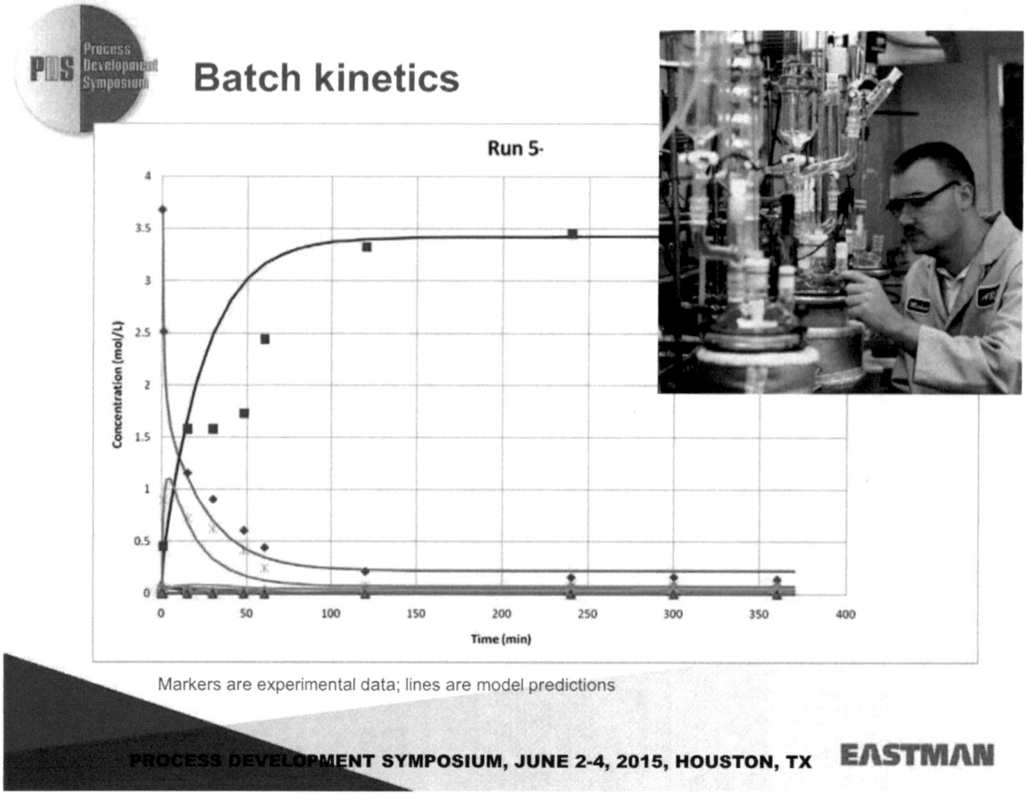

Process Development Symposium
Batch kinetics
Run 5
Concentration (mol/L)
Time (min)
0
0.5
1
1.5
2
2.5
3
3.5
4
0
50
100
150
200
250
300
350
400
Markers are experimental data; lines are model predictions
PROCESS DEVELOPMENT SYMPOSIUM, JUNE 2-4, 2015, HOUSTON, TX
EASTMAN

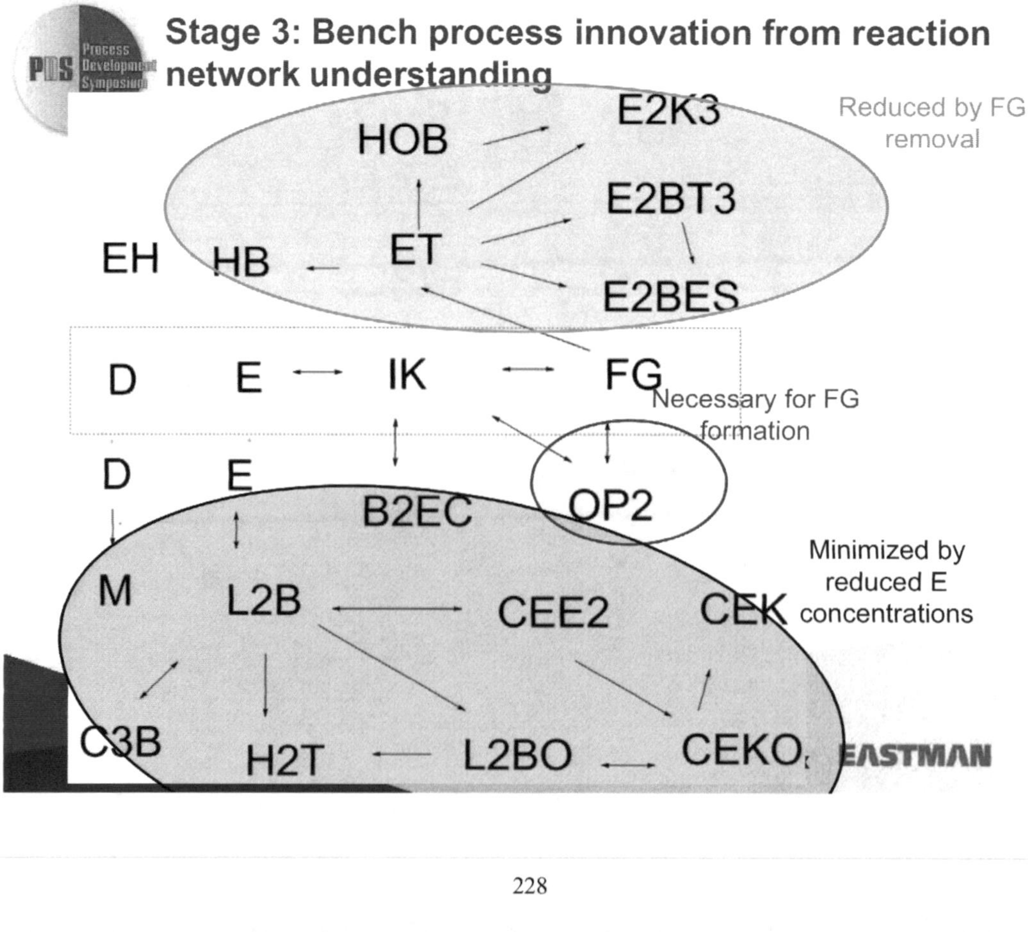

Process Development Symposium
Stage 3: Bench process innovation from reaction network understanding
Reduced by FG removal
HOB
E2K3
E2BT3
EH
HB
ET
E2BES
D
E
IK
FG
Necessary for FG formation
D
E
B2EC
OP2
Minimized by reduced E concentrations
M
L2B
CEE2
CEK
C3B
H2T
L2BO
CEKO
EASTMAN

Initial design: commercial and mini-plant

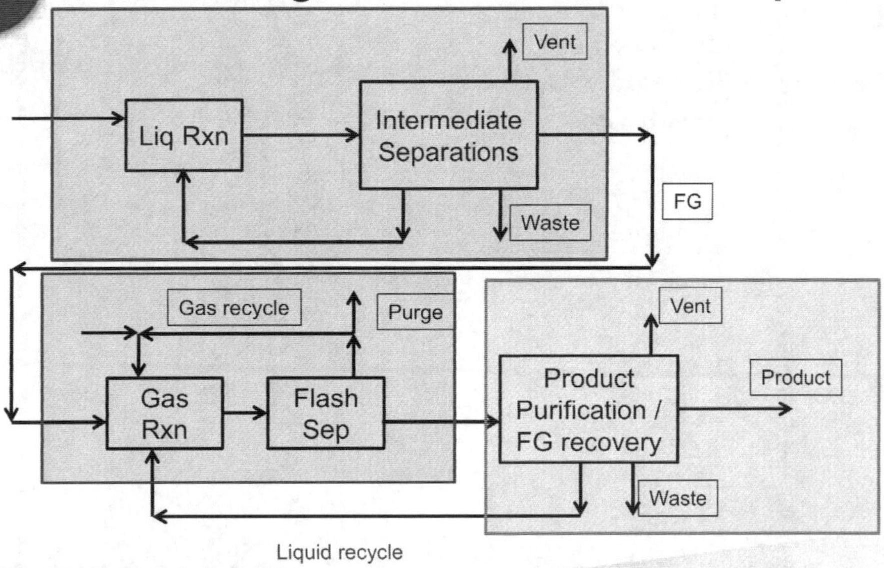

PROCESS DEVELOPMENT SYMPOSIUM, JUNE 2-4, 2015, HOUSTON, TX EASTMAN

Stage 4: Miniplant demonstration

- Initial commercial process design was completed
- Scaled down by > 100,000:1 ratio (~2 gr/min)
- Miniplant constructed to commercial design
- Process ran for 2½ months with only minor shutdowns
- Design rates & in-spec product achieved in 2 days
- All five loops closed
- All unit operations included
- No loss of catalyst activity or selectivity either reaction step
- Cost: < $2 MM (including design, purchase, installation, operation, analytical, process analysis, and documentation)
- Duration: 9 months (3 months design and construction, 3 months operations, 3 months analysis)

PROCESS DEVELOPMENT SYMPOSIUM, JUNE 2-4, 2015, HOUSTON, TX EASTMAN

First step

- High conversion
- High selectivity
- No evidence of catalyst deactivation

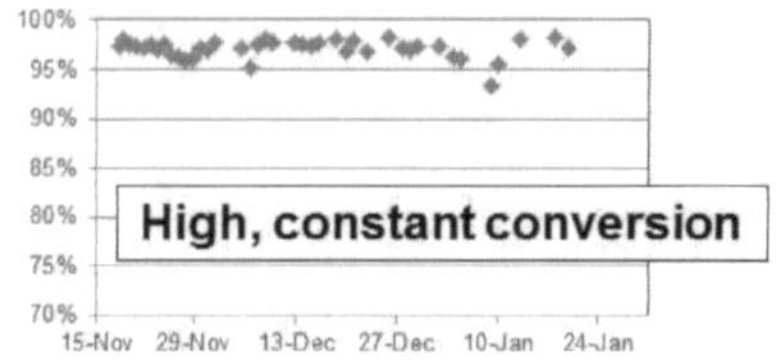

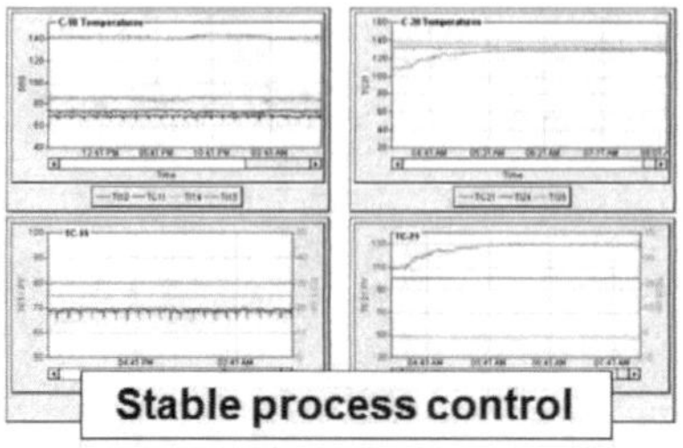

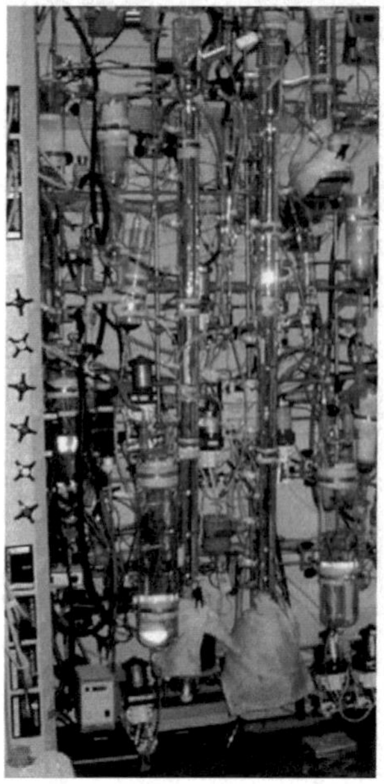

PROCESS DEVELOPMENT SYMPOSIUM, JUNE 2-4, 2015, HOUSTON, TX EASTMAN

Second step

Stable conversion and "C" selectivity observed over 71 days of operation

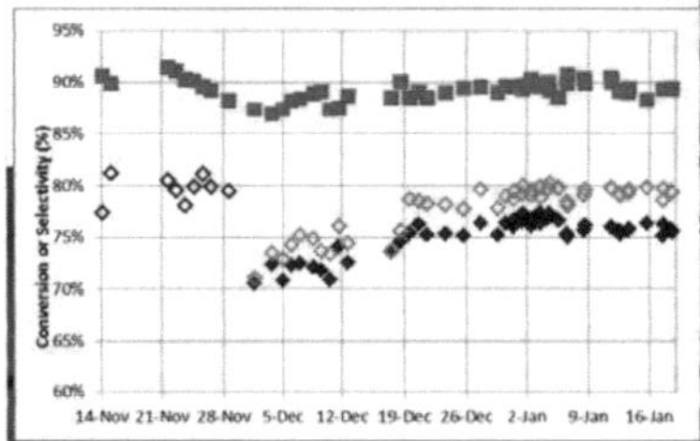

Catalyst post-mortem:
- Spent catalyst essentially indistinguishable from fresh

Corrosion studies reduced cost of MOC, saving $$ MM

PROCESS DEVELOPMENT SYMPOSIUM, JUNE 2-4, 2015, HOUSTON, TX EASTMAN

Product purification / FG recovery

- In-spec in 2 days and maintained
- All recycle loops closed

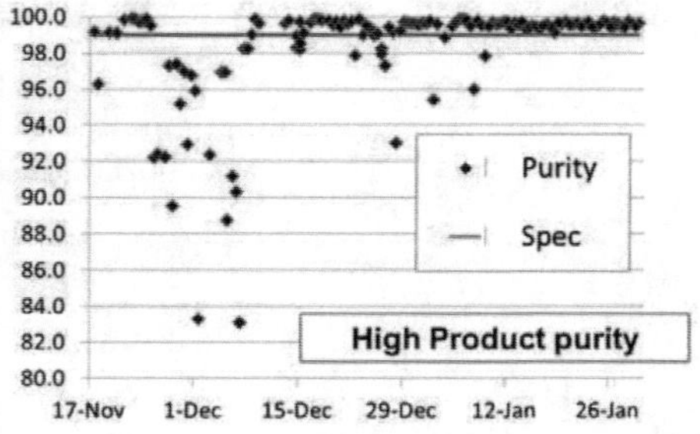

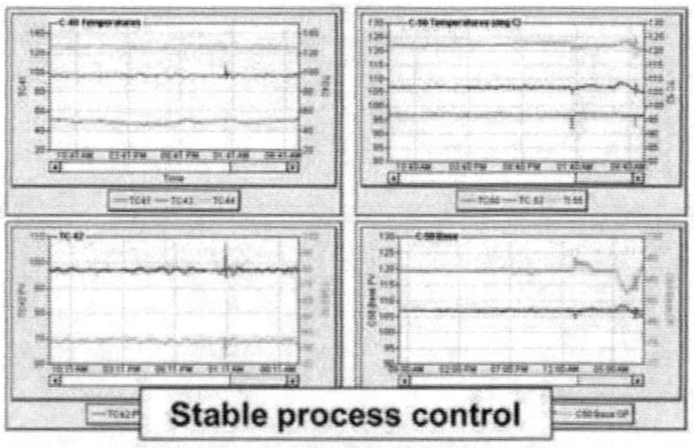

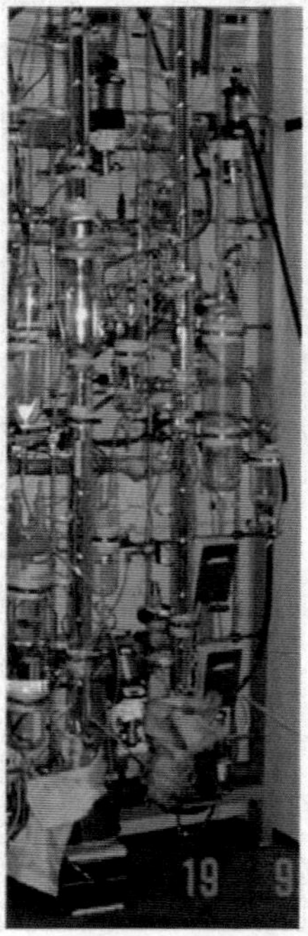

PROCESS DEVELOPMENT SYMPOSIUM, JUNE 2-4, 2015, HOUSTON, TX

EASTMAN

Preliminary Commercial Design

Preliminary commercial design
2nd step optimization

- Developed simplified HMB and economic model
- Used surrogate reactor model with iteration
- Tradeoff of system pressure, excess reactant, reactor & compressor size
- Final result – reduced section cost by 30% (model prediction 33% reduction)

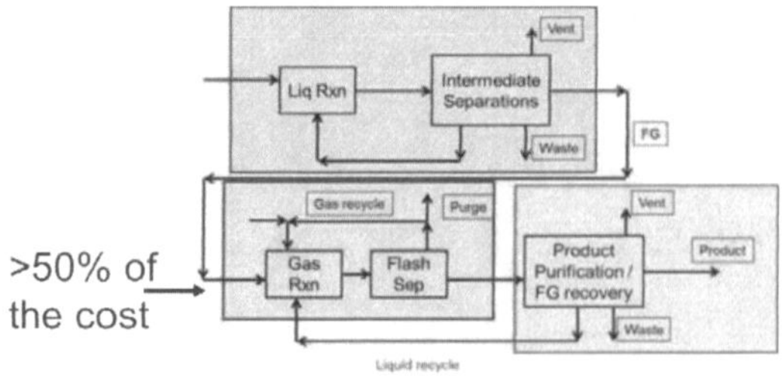

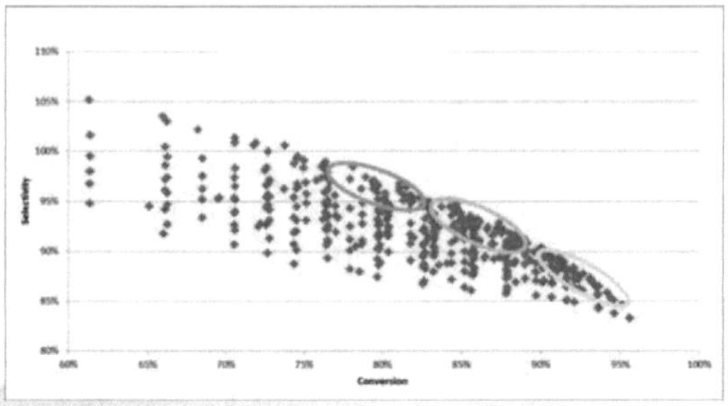

Preliminary Commercial Design
Materials of Construction

- Material of Construction selection was considered from the beginning of the project
- Corrosion coupons used in miniplant demonstration
- Lower cost MOC than was in original design found to be acceptable in most areas

PROCESS DEVELOPMENT SYMPOSIUM, JUNE 2-4, 2015, HOUSTON, TX EASTMAN

Preliminary Commercial Design Impact of Collaboration & Intentionality

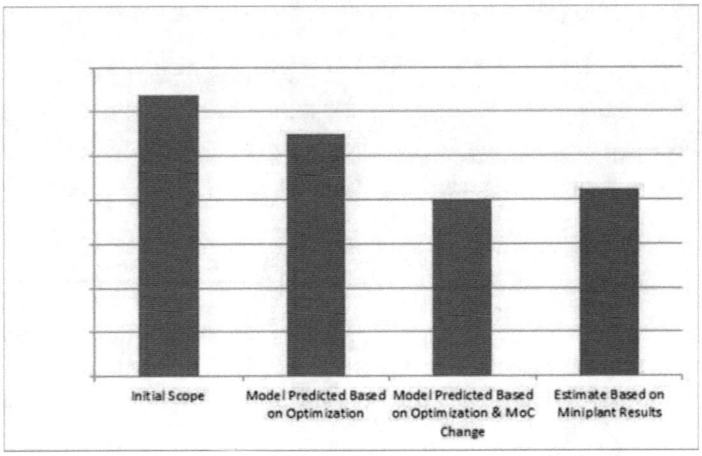

PROCESS DEVELOPMENT SYMPOSIUM, JUNE 2-4, 2015, HOUSTON, TX EASTMAN

Summary

- A 5-stage scale-up strategy was used to develop the process.
- Process innovation occurred when the reaction network was understood.
- Collaboration between Technology and Process Engineering began early in the project.
- Initial commercial design was scaled down by >100,000:1 ratio.
- Mini-plant demonstration operated for 2½ months with all unit operations and closed recycle loops.
- Collaboration led to capital reduction in preliminary commercial design.
- Deliverables for each stage were identified and can be completed by either Technology or Process Engineering.
- Collaboration and innovation led to a successful project.

PROCESS DEVELOPMENT SYMPOSIUM, JUNE 2-4, 2015, HOUSTON, TX EASTMAN

INVISTA™

Successful Scale-up & Commercialization of INVISTA's new Adiponitrile Technology

Benjamin (Ben) D. Herzog*, Larry Moerbe, Thomas Vos, Sudhir Aki, William Tenn, James McIntosh, Gerald Grunewald, Mark McPherson, and Gary Smith

INVISTA, 4123 East 37th Street North, Wichita, KS 67220

*ben.d.herzog@invista.com

Outline

- Importance of ADN
- INVISTA's ADN Chemistry
- Initial commercialization and implications
- Scale-up risk management
- Role of pilot plant; design considerations
- Commercial performance vs. pilot plant
- Summary

Importance of Adiponitrile

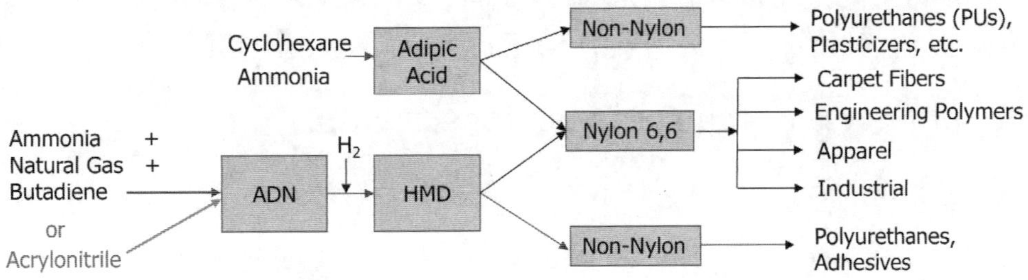

ADN: Adiponitrile
HMD: Hexamethylene diamine

- Adiponitrile (ADN) is a key intermediate for Nylon-6,6.
- ADN is hydrogenated to produce hexamethylene diamine (HMD) which is used to produce Nylon-6,6 and in other non-nylon applications such as coatings, adhesives, etc.

INVISTA

INVISTA ADN Synthesis

2-step hydrocyanation process

- Anti-Markovnikov addition of hydrogen cyanide to both ends of butadiene using a homogeneous catalyst.

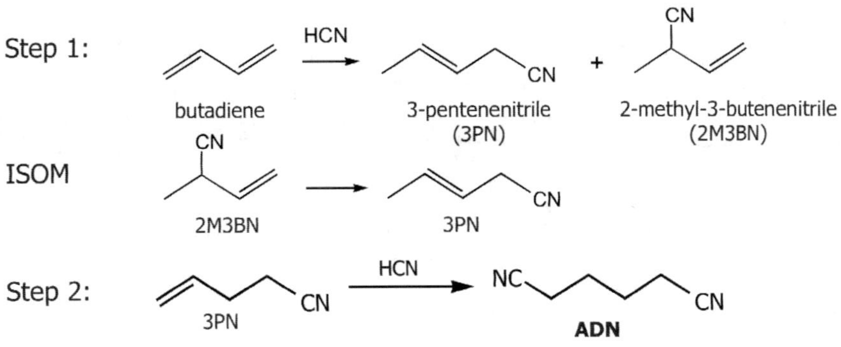

Pilot plant implication: significant recycle loops to close

INVISTA

Initial Commercialization & Implications

The first deployment of INVISTA's new ADN technology was a retrofit of the existing Orange, Texas ADN facility

Implications or requirements for scale-up:

- Technology works
- Able to monetize products
- Able to implement the new technology
- Risk adjusted project economics are attractive even in a downside case

The pilot plant was critical to address all of these requirements

Scale-up Risk Management

- Understanding, managing, and reducing risks were important drivers of the pilot unit, laboratory units, other lab based studies, and commercial demonstration programs
- The risks for the technology development, commercial scale-up, plant transition from the old to the new technology, and start-up were identified and assessed early in the program. These risks were input into a "failure mode and effects assessment" matrix. The impact and probability of the key risks occurring were quantified and challenged by a broad team from R&D, technology, operations, engineering, and commercial
- A key risks matrix was created and updated on a continuous basis by the development and commercialization teams
- Stress and downside cases were carefully defined for the key risks, quantified, and reviewed with senior leadership
- The team focused on reducing the probability of the stress and downside cases from occurring and minimizing the magnitude of the impact if the stress or downside case occurs

Conceptual Risk Matrix

Required to proceed with the retrofit	What related statements must be true	Success criteria (technical & financial)	Current status	Status at retrofit decision
We are confident in our ability to secure an adequate catalyst supply that meets our specification at a viable cost with an on-schedule timing to support the retrofit	Catalyst production and quality have been demonstrated at sufficient scale. Catalyst quality has been demonstrated in the pilot plant.	1. Achieve yield & productivity values that are consistent with our cost estimates. 2. Demonstrated process is robust 3. Demonstration is run at a scale that we believe is representative of reasonable scale risk 4. Catalyst manufacturing process is finalized	Process has been demonstrated	Process has been demonstrated

7

INVISTA

Role of the Pilot Plant

Demonstrate the technology at a scale to provide confidence in scaling the next step to commercial scale

- Confirm flowsheet conditions over an extended time
- Test steady state operation
  - Modelling was used to assess approach to steady state and impact of sampling
- Test operating window including start-up and shutdown, process upsets, etc.
- Close recycle loops
- Prototypic demonstration of unit operations
- Provide adequate quantities of materials for product demonstration, waste characterization, off-line testing, etc.
- Train operating personnel

8

INVISTA

Pilot Plant Design Approach and Considerations

- Flowsheet was simulated using ASPEN based on laboratory data
  - Flowsheet model was used extensively to design the pilot plant
- Parameters that influenced scale included:
  - Requirements associated with HCN as a feed
  - Demonstration needs on previous slide
  - Cost—capital and operating cost
  - Staffing/Timing
- Modular construction was used

INVISTA™

Commercial Performance vs. Pilot Plant

- Performance in the Orange plant matched pilot plant operation

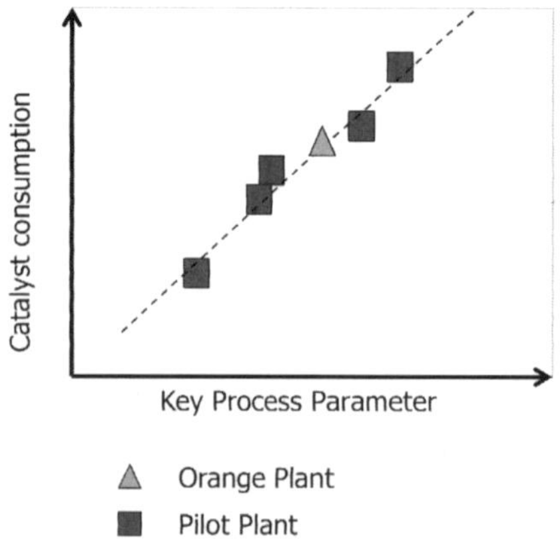

INVISTA™

Summary

- New ADN technology was successfully started up
- Pilot plant was critical to the successful commercialization
  - Pilot plant designed and sized to address key risks of scale-up
  - Use of pilot plant to demonstrate steady state and non-steady state conditions
  - Prototypical demonstration of unit operations in pilot plant
  - Full closure of all recycles

INVISTA™

AUTHOR INDEX